OXFORD LOGIC GUIDES

GENERAL EDITOR : DANA SCOTT

BEGINNING MODEL THEORY

THE COMPLETENESS THEOREM AND SOME CONSEQUENCES

BY

JANE BRIDGE

CLARENDON PRESS · OXFORD

1977

Oxford University Press, Walton Street, Oxford OX2 6DP

OXFORD LONDON GLASGOW NEW YORK
TORONTO MELBOURNE WELLINGTON CAPE TOWN
IBADAN NAIROBI DAR ES SALAAM LUSAKA ADDIS ABABA
KUALA LUMPUR SINGAPORE JAKARTA HONG KONG TOKYO
DELHI BOMBAY CALCUTTA MADRAS KARACHI

ISBN 0 19 853157 5

Printed in Great Britain
by Morrison & Gibb Limited, Edinburgh

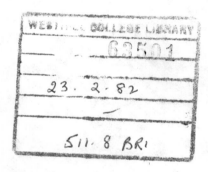

PREFACE

An essential feature of mathematical logic distinguishing it from other branches of mathematics, is the explicit study of the *language* in which mathematics may be described. This book is intended as an introduction to that aspect of logic which concerns the relationship between, on the one hand, mathematical *theories*, described in a formal language, and on the other, mathematical *structures* realizing those theories.

The text is based on a series of lectures given in Oxford each year from 1970 to 1973. The course was intended for second-year undergraduates reading Mathematics and Philosophy who had already had an introductory logic course. As a consequence, we assume here familiarity with some elementary concepts and results. For instance, the reader should ideally have studied the propositional (sentential) calculus, including the notions of a propositional language, a tautology, and a formal system of axioms and rules for generating all tautologies as theorems.† Acquaintance with the formalization of statements involving quantifiers is desirable though not strictly essential. Mathematical examples are given for which some familiarity with elementary abstract algebra is necessary. A summary of some set-theoretic concepts used in the text is given in the introduction.

At this point a brief summary outlining the contents of the book may be helpful. In Chapter 1 we consider the *objects* with which the mathematician is primarily concerned:

---

†Mendelson (1964), Chapter 1, or Lemmon (1965), Chapters 1 and 2, contain the relevant material. Scott's book in this series will cover the Oxford introductory course.

mathematical structures. A formal (first-order) language is developed in which some of the properties of a particular structure or class of structures may be expressed. The formalizable properties include many of the attributes of mathematical interest. (Such attributes are normally expressed in English, by English-speaking mathematicians, together with some standard abbreviations.) The grammar of the language is described and a definition given of what it is for a statement of the language to be *true* or *false* in a given structure.

Following the discussion of mathematical structures and the (first order) languages in which their properties can be formulated we turn, in Chapter 2, to the 'syntactic' notions. We define a formal system comprising axioms, rules of proof, and derivations and develop some elementary theory of the system.

Chapter 3 is the core of the book; the semantics of Chapter 1 (which concern the 'real' mathematical world and meaning in that world) are related to the syntactic (purely formal) ideas of Chapter 2. The formal system codes an intuitive idea of 'provable from certain axioms'; moreover the system is strong enough to generate all intuitively acceptable 'theorems' (expressible in the system). This important result, *Gödel's completeness theorem* is the first substantial theorem of the book and, together with its corollaries, is the essential foundation for nearly all further work in model theory.

The final chapter is a selective introduction to topics in what might be termed 'proper model theory'. The directions taken are by no means the only ones which properly follow the completeness theorem. The aim is to convey the flavour of the subject and to interest the reader sufficiently so that he will be encouraged to study further with the help of one of the more advanced standard texts on model theory.

I would like to thank several people for their help during the preparation of this book. Robin Gandy, Anne Troelstra, Derek Goldrei, and Michael Katz each read an early version of the text and provided much constructive criticism and advice. At a later stage a duplicated version of the text was used in Oxford and several students pointed out to me errors and inaccuracies. I am grateful to them and to Alan Slomson who suggested further corrections and improvements. Finally I am indebted to Dana Scott whose idea it was originally that I should write this book. His help during the preparation has been invaluable.

October 1975 J.B.

CONTENTS

INTRODUCTION
BACKGROUND SET THEORY

The fundamental mathematical concepts to be considered in this book can only be defined relative to more basic, and it is hoped simpler, more intuitive concepts. The concepts we accept as our starting point are those of (naive) set theory. The student having only minimal acquaintance with text books of pure mathematics will be familiar with the introductory chapter on set theory such books invariably contain. We will assume knowledge of the contents of such a chapter. In order to comprehend the most general form of the completeness theorem in Chapter 3 and some of its corollaries in Chapter 4 the axiom of choice and some simple cardinal arithmetic are needed. The following remarks highlight the aspects of set theory which are most important for our purposes.

A *set* is a collection of *elements*. When the element s is a member of the set S we write $s \in S$ and, conversely, $s \notin S$ denotes that s is not a member of S. Two sets S and T are the same just in case they have precisely the same elements (the *axiom of extensionality*). Hence S may be described (uniquely) by exhibiting those elements which are members of S. $S = \{x_1,\ldots,x_n\}$ denotes that S has as members precisely $x_1,\ldots,x_n$. More generally, $S = \{x : P(x)\}$ denotes that S consists of those elements x which satisfy the proposition $P(x)$. The set S is a *subset* of the set T, $S \subseteq T$, if every element of S is also an element of T. The *empty set*, the set with no members, is denoted by $\emptyset$. With any two elements a,b we associate the *ordered pair* $\langle a,b\rangle$ which has the following property: $\langle a,b\rangle = \langle a',b'\rangle$ just in case $a = a'$ and $b = b'$. (Any set-theoretic definition which has this property is sufficient for our purposes. One example in common use is $\langle a,b\rangle =_{Df} \{\{a\}, \{a,b\}\}$.) Now by induction on n for $n > 2$ the *ordered n-tuple* $\langle a_1,\ldots a_n\rangle$ of elements $a_1,\ldots,a_n$, is defined to be the ordered pair $\langle\langle a_1,\ldots a_{n-1}\rangle ,a_n\rangle$.

Given any two sets S and T the *union*, $S \cup T$, is $\{r: r \in S \text{ or } r \in T\}$; the *intersection*, $S \cap T$, is $\{r: r \in S \text{ and } r \in T\}$; and the *cartesian product*, $S \times T$, is $\{\langle s,t\rangle : s \in S, t \in T\}$. More generally, if $S_1,\ldots,S_n$ are sets and $n \geq 2$ then we define

$$S_1 \cup \ldots\ldots \cup S_n = \{s: s \in S_i \text{ for some } i=1,\ldots,n\},$$

$$S_1 \cap \ldots\ldots \cap S_n = \{s: s \in S_i \text{ for all } i=1,\ldots,n\}, \text{ and}$$

$$S_1 \times \ldots\ldots \times S_n = \{\langle s_1,\ldots,s_n\rangle : s \in S_i \text{ for } i=1,\ldots,n\}.$$

$S \times \ldots\ldots \times S$ (n times) is sometimes abbreviated by S^n.

An *n-ary relation* R $(n \geq 2)$ *on the set* S is a subset of S^n. A *unary relation on* S is simply a subset of S. An *n-ary function* f *on* S $(n \geq 1)$ is an $(n+1)$-ary relation on S with the property that for each n-tuple $\langle s_1,\ldots,s_n\rangle \in S^n$ there is a unique $s \in S$ (usually written $f(s_1,\ldots,s_n)$) such that $\langle s_1,\ldots,s_n,s\rangle \in f$. More generally, *a function* f *from* S *to* T written $f : S \to T$, is a subset of $S \times T$ such that for each $s \in S$ there is a unique $t \in T$ with $\langle s,t\rangle \in f$. When $\langle s,t\rangle \in f$ we write $f(s) = t$. If f is a function from S to T *the domain of* f is S and the *range of* f is $\{t \in T: \text{for some } s \in S, \langle s,t\rangle \in f\}$. f is *onto* when the range of f is the whole of T and *one-one* when for each t in the range of f there is a unique $s \in S$ with $\langle s,t\rangle \in f$. f is a *bijection* when it is one-one and onto.

Suppose $\{S_i: i \in I\}$ is a collection of sets indexed by elements of the set I. The *union* of $\{S_i: i \in I\}$, $\bigcup_{i\in I} S_i$, is $\{s: s \in S_i \text{ for some } i \in I\}$ and the *intersection* of $\{S_i: i \in I\}$ is $\{s: s \in S_i \text{ for all } i \in I\}$. The *product* $\bigtimes_{i\in I} S_i$ is defined to be $\{f: f \text{ is a function, } f: I \to \bigcup_{i\in I} S_i$ and $f(i) \in S_i \text{ for all } i \in I\}$. (When I is $\{1,\ldots,n\}$ then we associate $f: I \to \bigcup_{i=1,\ldots,n} S_i$ with $\langle f(1),\ldots,f(n)\rangle$ to accord with our earlier definition of $S_1 \times \ldots\ldots \times S_n$.) The *axiom of choice* states that if for each $i \in I$ the set S_i is a non-empty set then the product $\bigtimes_{i\in I} S_i$ is also non-empty. The axiom implies that there is at least one function, a *choice function* which picks out from each of a family of non-empty sets an element of that set. There are several important consequences of this axiom and essential use of one of them, the well ordering theorem, will be made in Chapter 3. First, a few more definitions are necessary. A binary relation R on S is a *partial order* on S if

(i) for every $s \in S$, $\langle s,s\rangle \in R$ (R is reflexive),

(ii) for every $s_1,s_2,s_3 \in S$ if $\langle s_1,s_2\rangle \in R$ and $\langle s_2,s_3\rangle \in R$ then $\langle s_1,s_3\rangle \in R$ (R is transitive),

and *(iii)* for every $s_1,s_2 \in S$ if $\langle s_1,s_2\rangle \in R$ and $\langle s_2,s_1\rangle \in R$ then $s_1 = s_2$ (R is anti-symmetric).

When R is a partial order on S then we write $s_1 \leq s_2$ for $\langle s_1,s_2\rangle \in R$. The ordered pair $\langle S,\leq\rangle$ denotes the partially ordered set. The partially ordered set $\langle S,\leq\rangle$ is *totally ordered* if for every $s_1,s_2 \in S$ either $s_1 \leq s_2$ or $s_2 \leq s_1$ ($\leq$ is connected). A *well order* $\leq$ on S is a total order such that for every non-empty $S_0 \subseteq S$ there is an $s_0 \in S_0$ such that $s_0 \leq s$ for all $s \in S_0$ (every non-empty subset of S has a minimal element).

The *well ordering theorem*, which both implies and is a consequence of the axiom of choice, states that every non-empty set can be well ordered. An alternative equivalent of the axiom of choice which will not be used in the text but is referred to in the exercises is *Zorn's lemma*: suppose $\langle X,\leq\rangle$ is a non-empty partially ordered set such that every totally ordered subset $Y \subseteq X$ has an upper bound in X (i.e. there is an element $x_Y \in X$ such that for all $y \in Y, y \leq x_Y$); then X has a maximal element (i.e. an element $m \in X$ such that for all $x \in X, m \leq x$ implies $m = x$).

A binary relation $<$ is a *strict partial* [*total*] *order* on S if

(i) for every $s \in S$, not $s < s$ (irreflexive)

(ii) for every $s_1,s_2,s_3 \in S$ if $s_1 < s_2$ and $s_2 < s_3$ then $s_1 < s_3$ (transitive),

[*(iii)* for every $s_1, s_2 \in S$ either $s_1 < s_2$ or $s_1 = s_2$ or $s_2 < s_1$ (connected)].

A *strict well order* on S is a strict total order $<$ on S such that for each non-empty $S_0 \subseteq S$ there is an $s_0 \in S_0$ such that for each $s \in S_0$ either $s = s_0$ or $s_0 < s$. Any partial [total, well] order on S, say $\leq$, induces a strict partial [total, well] order of S, $<$, given by

$$s_1 < s_2 \quad \text{iff}^{\dagger} \quad s_1 \leq s_2 \quad \text{and} \quad s_1 \neq s_2.$$

Similarly given a strict partial order $<$ the relation $\leq$ defined by

$$s_1 \leq s_2 \quad \text{iff} \quad s_1 < s_2 \text{ or } s_1 = s_2$$

is a partial order.

A *transitive set* S is a set such that for all s_1, s_2 if $s_1 \in s_2 \in S$ then $s_1 \in S$ (or, equivalently, $s_2 \in S$ implies $s_2 \subseteq S$). A transitive set on which $\in$ is a strict well order is an *ordinal*. According to this definition the empty set is an ordinal (in this context usually written as 0) and if α is an ordinal then $\alpha \cup \{\alpha\}$ is an ordinal, the *successor* of α, sometimes written as $\alpha + 1$. An ordinal β which is not the successor of an ordinal and is not 0 is a *limit* ordinal and has the property $\beta = \bigcup\beta = \{\alpha : \alpha \in \gamma \text{ for some } \gamma \in \beta\}$. The least limit ordinal, $\{\emptyset, \{\emptyset\}, \{\emptyset, \{\emptyset\}\}, \ldots\}$, is denoted by ω. Since $\in$ is a strict well order on any set of ordinals we sometimes write $\alpha < \beta$ for α, β ordinals and $\alpha \in \beta$. Every well ordered set $\langle X, \leq\rangle$ is order-isomorphic to a unique ordinal; that is, there is an ordinal α and a one-one onto function $f\colon X \to \alpha$ such that if $x_1 \leq x_2$ then $f(x_1) \in f(x_2)$ or $f(x_1) = f(x_2)$. So the well ordered set $\langle X, \leq\rangle$ can be written $\{x_\beta : \beta \in \alpha\}$ where x_β is the unique inverse image of β under f. The principle of mathematical induction for the natural numbers can be extended to the *principle of transfinite induction* for the ordinals (or, indeed, any well ordered set):

---

† Throughout 'iff' will be used as an abbreviation for 'if and only if'.

suppose $P(\alpha)$ is a property of ordinals such that for any ordinal β, when $P(\gamma)$ holds for all ordinals $\gamma < \beta$ then $P(\beta)$ holds; then $P(\alpha)$ holds for all ordinals α.

The *cardinal* of a set X is denoted by $\overline{X}$ and is a measure of the size of X. If X and Y are sets then we define $\overline{X} \le \overline{Y}$ (Y is at least as large as X) if there is a one-one function $f\colon X \to Y$; $\overline{X} = \overline{Y}$ if $\overline{X} \le \overline{Y}$ and $\overline{Y} \le \overline{X}$ or, equivalently,[†] if there is some one-one and onto function $f\colon X \to Y$. A corollary of the axiom of choice is that for any two sets X,Y either $\overline{X} \le \overline{Y}$ or $\overline{Y} \le \overline{X}$; moreover cardinals may be identified with certain *initial ordinals* (defined as those which cannot be put in one-one correspondence with strictly lesser ordinals). Let N denote the set of all natural numbers (with 0) and $\aleph_0 = \overline{N}$. A set X is *finite* iff for some natural number n,X is in one-one correspondence with $\{0,\ldots,n-1\}$. X is *infinite* if it is not finite. X is *countable* if $\overline{X} \le \aleph_0$. Suppose $\overline{X} = \kappa$ and $\overline{Y} = \lambda$; the *cardinal sum* $\kappa + \lambda$ is defined to be $\overline{X^* \cup Y^*}$ where X^*,Y^* are disjoint sets with the same cardinal as X,Y respectively (e.g. $X^* = X \times \{0\}$, $Y^* = Y \times \{\{0\}\}$); the *cardinal product* $\kappa\lambda$ is $\overline{X \times Y}$; and *cardinal exponentiation* κ^λ is defined to be $\overline{X^Y}$ where $X^Y = \{f\colon f \text{ is a function from } Y \text{ into } X\}$. Some simple laws of cardinal arithmetic, which follow from the assumption of the axiom of choice, are stated in Chapter 3 (lemma 3.17). Full details of all these set-theoretic concepts may be found in any standard text book of set theory, e.g. Halmos (1960).

[†]By the Schröder-Bernstein theorem; see Halmos (1960) page 88.

1
RELATIONAL STRUCTURES

1.1. DEFINITIONS AND EXAMPLES

Mathematics is rich in structures. Euclid's notion of three-dimensional space with points, lines, and planes and the more recent notion of a group may be construed as typical examples of a concept of 'abstract structure' that is fundamental to the mathematician. Loosely speaking, such a structure consists of a collection of objects, (first order) relations between the objects, (second-order) relations between those relations, and so on. The structure is 'abstract' in the sense that the nature of the objects and of the relations is, mathematically speaking, irrelevant; the importance of the structure from this point of view lies in the underlying pattern of the objects together with the relations.† Euclidean space is a structure whose objects are points, lines, and planes; relations are those such as incidence, collinearity, and parallelism. A group is a structure in which the key relation is that determined by the group operation. Our first definition makes explicit a particular form of this abstract concept, namely that of a first order structure. This set-theoretic foundation will underlie our study of model theory.

<u>Definition 1.1.</u> A *relational structure* is an ordered quadruple

$$\mathfrak{A} = \langle A, \{R_i\}_{i \in I}, \{f_j\}_{j \in J}, \{c_k\}_{k \in K} \rangle \quad \text{with}$$

associated functions $\lambda : I \to N^+$, $\mu : J \to N^+$

such that

---

†See Gandy (1973) for a lucid essay on 'structure' as it occurs in mathematics.

(i) A, the *domain* of $\mathfrak{A}$, is a non-empty set,

(ii) I is a (possibly empty) set such that for each $i \in I$ $\lambda(i)$ is a positive integer and R_i is a $\lambda(i)$-ary relation on A,

(iii) J is a (possibly empty) set such that for each $j \in J$ $\mu(j)$ is a positive integer and f_j is a $\mu(j)$-ary function on A,†

(iv) K is a (possibly empty) set such that for each $k \in K$ c_k, a *distinguished element* (or *constant*), is an element of A.

As indicated in the definition we use a Gothic letter for a relational structure and adopt the convention that the corresponding Roman letter denotes the domain of the structure. If we allow A to be empty then certain technical problems arise from subsequent definitions depending on the notion of a relational structure. These can be overcome by considering separately the case where A is empty. However we prefer to eliminate this case at the outset. I, J, K are indexing sets for the relations, functions, and distinguished elements respectively. λ,μ are functions from I, J respectively to N^+, the set of positive integers, coding the arity of the relations and functions. We recall that unary relations on A are identified with subsets of the domain A.

This unwieldy definition covers a very wide family of mathematical structures. (Note however that we have excluded structures with proper classes as domain.) For many of the familiar mathematical structures a more restricted but simpler definition would have sufficed. In some instances, for example, the sets J and K are empty. When this occurs

---

†Most set theoretic definitions of 'function' identify it with a certain type of relation, a so-called functional relation (see page 2). Here we wish to distinguish between some such relations, indexed by elements of J, and arbitrary relations, indexed by elements of I.

the relational structure is effectively an ordered pair consisting of a non-empty domain A and a set of relations on A; it is then said to be purely relational. Other simple forms of structure are exemplified below.

Example 1.2. A non-empty set S can be considered as a purely relational structure with a single binary relation, =, the equality relation, $\langle S,\{=\}\rangle$.

When, as in this example, the set I is finite we omit the brackets { } and instead list the relations. Thus the example $\langle S,\{=\}\rangle$ is written $\langle S,=\rangle$ and, in general, if I has n elements $\{R_i\}^{\mathfrak{A}}_{i\in I}$ is replaced by $R_1,\ldots,R_n$. A similar replacement is made if J or K is finite.

To include the equality relation in the presentation of a structure is, in a sense, superfluous. It does not imply any 'structure' on the underlying domain. However, in section 1.2 we introduce a language associated with a given structure and this depends intrinsically on the relations, functions, and constants explicitly displayed in the presentation. Two different languages arise depending on whether or not the equality relation is included. Most mathematical languages do include a symbol for equality. At this stage though we do not assume that the equality relation is given in the presentation of a relational structure.

Example 1.3. The ordered set of natural numbers N with least element 0 is a structure $\langle N,<,0\rangle$ with one binary relation, <, such that the pair $\langle n,m\rangle \in <$ just in case $n < m$ in the usual ordering.

Some familiar mathematical examples have a number of different characterizations as relational structures.

Example 1.4.1. A group may be presented as a relational structure of the form $\langle G,=,\cdot,e\rangle$, where G is the set of elements of the group, = is the equality relation on G, $\cdot$, a binary function, is the group operation, and e is the identity element. We could also include a unary function, $^{-1}$, representing the inverse operation.

Example 1.4.2. A group may also be presented as a structure with the following form: $\mathfrak{G} = \langle G, M, e\rangle$ where M is a ternary relation representing the group operation '·' in the following way:

$$\langle g_1, g_2, g_3\rangle \in M \text{ iff } g_1 \cdot g_2 = g_3.^{\dagger}$$

The group axioms then ensure that

(i) if $\langle g_1, g_2, g_3\rangle, \langle g_2, g_4, g_5\rangle \in M$ then there is a unique $g_6 \in G$ such that $\langle g_1, g_5, g_6\rangle$, $\langle g_3, g_4, g_6\rangle \in M$ (associativity of the group operation);

(ii) for each $g \in G$, $\langle g, e, g\rangle \in M$ and $\langle e, g, g\rangle \in M$ (e is the identity element);

and (iii) for each $g \in G$ there is a unique $g^{-1} \in G$ such that $\langle g, g^{-1}, e\rangle, \langle g^{-1}, g, e\rangle \in M$ (every element has an inverse).

Alternative presentations may be given either by omitting the constant e or by adding the equality relation. The structure $\langle G, M, e\rangle$ of example 1.4.2 is, from a purely formal point of view, simpler than the structure $\langle G, =, \cdot, e\rangle$ of example 1.4.1. The function $\cdot$ is replaced by the relation M. This kind of replacement is always possible when the set of functions $\{f_j\}_{j\in J}$ in the relational structure $\mathfrak{A}$ is non-empty. We substitute for the $\mu(j)$-ary function f_j the $(\mu(j)+1)$-ary relation R_{f_j} given by

---

† If a function is defined as a certain kind of relation (as on page 2) then M is just $\cdot$! However the presentation differs from that in 1.4.1 because J is now empty and there is a single ternary relation in place of a binary function.

$$\langle a_0,\ldots,a_{\mu(j)}\rangle \in R_{f_j} \text{ iff } f_j(a_0,\ldots,a_{\mu(j)-1}) = a_{\mu(j)}.$$

We can go one stage further and replace each distinguished constant c by the unary relation $R_c = \{c\}$ so that the resulting structure $\mathfrak{A}^R = \langle A, \{R_i\}_{i\in I} \cup \{R_{f_j}\}_{j\in J} \cup \{R_{c_k}\}_{k\in K}\rangle$ is purely relational. Although purely relational structures are technically simpler than the most general kind of relational structure we prefer, in most cases, to consider structures with the added complexity due to functions and constants. In the transition from a function f to the associated relation R_f the functional character is obscured. Although functions are sometimes introduced as special kinds of relations they themselves are so fundamental in mathematics that it is natural to exhibit them explicitly as functions in the structure. It is for instance more natural to think of a group as a set G together with an operation $\cdot : G \times G \to G$ and a distinguished element, e, the identity, rather than as a set G with a ternary relation R and a unary relation R_e. From a mathematical viewpoint the two structures $\mathfrak{A}$ and the associated purely relational structure $\mathfrak{A}^R$ are identical. However as relational structures they are distinct. For the remainder of this book a relational structure is assumed to have a unique presentation of the form

$$\langle A, \{R_i\}_{i\in I}, \{f_j\}_{j\in J}, \{c_k\}_{k\in K}\rangle.$$

Exercise 1.1. Give at least two distinct presentations of a field as a relational structure.

The final example of a relational structure demonstrates that the definition allows the inclusion of structures with more than one kind of basic element.

Example 1.5. A vector space over a field F is a structure

$$\mathfrak{B} = \langle V, U, F, +, \cdot, 0_U, 0_F, 1_F\rangle$$

where V is the set of vectors and scalars (elements of F), U, F are unary relations such that

$$v \in U \text{ iff } v \text{ is a vector,}$$

and $$\alpha \in F \text{ iff } \alpha \text{ is a scalar,}$$

0_U is the zero vector, 0_F is the zero scalar, 1_F is the unit element of F, + is a binary function representing both scalar and vector addition, and $\cdot$ is a binary function representing scalar multiplication and multiplication of a vector by a scalar. In the definition of a relational structure functions are assumed to be defined on the whole domain. However in an arbitrary vector space multiplication of two vectors is not defined. Likewise there is no operation of addition on a vector and a scalar. In order to construe a vector space as a structure in the sense of definition 1.1 we suppose that multiplication is defined on a pair of vectors in some purely arbitrary way; for example $v_1 \cdot v_2 = 0_U$ for any vectors v_1, v_2. Similarly if $\alpha \in F$ and v is a vector then $\alpha + v$ is given some arbitrary value (in V). Thus in a sense we have imposed extra structure on the vector space so as to bring it in line with the general definition. If the corresponding ternary relations $R_\cdot$ and R_+ are used in place of the functions $\cdot$ and + then no such problem occurs.†

---

†The example given of a vector space is just one instance of a common phenomenon in mathematics, that of a structure with several different sorts of basic objects. Another familiar example is Euclidean space with points, lines, and planes. Although such examples can be construed as relational structures of the type defined in 1.1 (as is done above for vector spaces) an alternative method is to consider many-sorted structures (i.e. structures with an indexed set of domains each containing objects of a different sort) from the start. For further details of this approach see e.g. Kreisel-Krivine (1967) Chapter 5, or Wang (1952).

Because the class of relational structures is so wide it is convenient to be able to consider those structures in some well-defined subclass. The following definition provides a useful classification.

Definition 1.6. Given a relational structure

$$\mathfrak{A} = \langle A, \{R_i\}_{i \in I}, \{f_j\}_{j \in J}, \{c_k\}_{k \in K} \rangle$$

the *type of* $\mathfrak{A}$, $\delta_{\mathfrak{A}}$, is the ordered triple

$$\langle K, \lambda, \mu \rangle$$

where $\lambda: I \to N^+$, $\mu: J \to N^+$ are functions coding the arity of the relations and functions respectively. The type of a structure is simply a measure of the relations, functions (together with their arities) and constants in the structure. Notice that is does *not* code the cardinality of the domain. Two structures $\mathfrak{A}$ and $\mathfrak{B}$ are of the *same type* just in case $\delta_{\mathfrak{A}}$ is the same as $\delta_{\mathfrak{B}}$. Two structures may be of the same type even when the domains are of different cardinalities but their relations, functions, and constants will be in one-one correspondence. Two groups presented as $\langle G, =, \cdot, e \rangle$ and $\langle G', =, +, 0 \rangle$ are of the same type according to definition 1.6 but the groups $\langle G, =, \cdot, e \rangle$ and $\langle G', A, 0 \rangle$ (where A is the ternary relation representing +) are not. We say the structures $\mathfrak{A}$ and $\mathfrak{B}$ are of *similar type* if $\delta_{\mathfrak{A}} = \langle K, \lambda, \mu \rangle$ and $\delta_{\mathfrak{B}} = \langle K', \lambda', \mu' \rangle$ and there are bijections $h_1 : I \to I'$, $h_2 : J \to J'$, $h_3 : K \to K'$ (I', J' are the domains of λ', μ' respectively) such that $\lambda(i) = \lambda'(h_1(i))$ for each $i \in I$ and $\mu(j) = \mu'(h_2(j))$ for each $j \in J$. Thus structures of similar type can readily be re-indexed so as to become structures of the same type.

It is usual to single out from an arbitrary algebraic structure certain distinguished subsets. We consider subgroups of groups, ideals and subrings of rings, subspaces of vector spaces, and so on. Such subsets are characterized as those non-empty subsets which are closed under the algebraic operations defined on the structure. These are themselves structures under the induced relations and restricted functions. Consideration of this construction motivates the following definition.

Definition 1.7. Suppose $\mathfrak{A} = \langle A, \{R_i\}_{i\in I}, \{f_j\}_{j\in J}, \{c_k\}_{k\in K} \rangle$. A *substructure* of $\mathfrak{A}$ is a relational structure of the same type as $\mathfrak{A}$, say $\mathfrak{B} = \langle B, \{S_i\}_{i\in I}, \{g_j\}_{j\in J}, \{d_k\}_{k\in K} \rangle$ satisfying the following conditions:

(i) $B \subseteq A$ (the domain of a substructure of $\mathfrak{A}$ is a subset of the domain of $\mathfrak{A}$);

(ii) for each $i \in I$, $R_i \cap B^{\lambda(i)} = S_i$ (S_i is the restriction of R_i to the domain of $\mathfrak{B}$);

(iii) for each $b_1,..,b_{\mu(j)} \in B$, each $j \in J$, $g_j(b_1,..,b_{\mu(j)}) = f_j(b_1,..,b_{\mu(j)})$ (g_j is the restriction of the function f_j to the domain of $\mathfrak{B}$ which must therefore by definition 1.1 be closed under the function f_j);

(iv) for each $k \in K$, $c_k = d_k$ (each distinguished element of $\mathfrak{A}$ is in the domain of a substructure of $\mathfrak{A}$).

If $\mathfrak{B}$ is a substructure of $\mathfrak{A}$ then $\mathfrak{A}$ is an *extension* of $\mathfrak{B}$. We write $\mathfrak{B} \subseteq \mathfrak{A}$ or $\mathfrak{A} \supseteq \mathfrak{B}$.

Exercise 1.2. Show that $\subseteq$ is a partial ordering on the class of realizations of type δ.

We now reconsider some of the earlier examples in the light of this definition. A substructure of the set $\langle S,=\rangle$ (example 1.2) is trivially a subset and conversely a subset is a substructure. A subgroup $\langle H,=,\cdot,e\rangle$ of the group $\langle G,=,\cdot,e\rangle$ (example 1.4.1) contains the distinguished element e and is closed under the operation $\cdot$; it is therefore a substructure. However a substructure $\langle G',=,\cdot,e\rangle$ of the group $\langle G,=,\cdot,e\rangle$ is not necessarily a subgroup. There is no reason to suppose that G' is closed under inverses. For example the non-zero integers form a substructure $\langle Z - \{0\}, =, \cdot, 1\rangle$ but not a subgroup of the non-zero reals under multiplication $\langle R - \{0\}, =, \cdot, 1\rangle$. Furthermore substructures of the group $\langle G, M, e\rangle$ (example 1.4.2), say $\langle G', M', e\rangle$, are not necessarily closed under the group operation, let alone under the inverse operation. Despite this it is possible to present a group in such a way that the substructures are precisely the subgroups. For consider a substructure $\langle H, =, \cdot, {}^{-1}, e\rangle$ of the group $\langle G, =, \cdot, {}^{-1}, e\rangle$. By definition 1.7 *(iv)* $e \in H$ and by *(iii)* H is closed under $\cdot, {}^{-1}$.

<u>Exercise 1.3.</u> Is a substructure of the ring

$$\langle R, =, +, \cdot, 0\rangle \quad \text{necessarily a subring?}$$

What happens if the unary additive inverse operation is added to the presentation of the ring?

<u>Exercise 1.4.</u> Give a presentation of a field so that the substructures are precisely the subfields.

Just as a subset S of a group G generates a subgroup G_S, the smallest subgroup of G containing S, so a subset S of the domain of a relational structure $\mathfrak{A}$ generates a substructure $\mathfrak{A}_S$, the smallest substructure of $\mathfrak{A}$ having a domain with S as subset. The domain of $\mathfrak{A}_S$ is the closure of $S \cup \{c_k\}_{k \in K}$ under the set of functions $\{f_j\}_{j \in J}$. Note that any non-empty subset of the domain of a purely relational structure $\mathfrak{A}$ is the domain of a substructure of $\mathfrak{A}$.

Mappings between two algebraic structures of the same kind that preserve structure are called homomorphisms. We adopt precisely the same terminology.

<u>Definition 1.8.</u> Suppose $\mathfrak{A} = \langle A, \{R_i\}_{i \in I}, \{f_j\}_{j \in J}, \{c_k\}_{k \in K}\rangle$ and $\mathfrak{B} = \langle B, \{S_i\}_{i \in I}, \{g_j\}_{j \in J}, \{d_k\}_{k \in K}\rangle$ are two structures of the same type. A map $h: A \to B$ is a *homomorphism* from $\mathfrak{A}$ to

$\mathfrak{B}$ iff for each $i \in I$, $j \in J$, $k \in K$, $a_1,\ldots,a_{\lambda(i)}$, $a_1,\ldots,a_{\mu(j)}, a \in A$,

$$\langle a_1,\ldots,a_{\lambda(i)}\rangle \in R_i \Rightarrow \langle h(a_1),\ldots,h(a_{\lambda(i)})\rangle \in S_i,$$

$$f_j(a_1,\ldots,a_{\mu(j)}) = a \Rightarrow g_j(h(a_1),\ldots,h(a_{\mu(j)})) = h(a),$$

and $$h(c_k) = d_k.$$

h is a mapping that preserves structure.
A homomorphism from $\langle G,=,\cdot,e\rangle$ to $\langle G',=,+,0\rangle$ when the two relational structures are groups is a group homomorphism in the usual group-theoretic sense.

Exercise 1.5. Is a homomorphism (of relational structures) from the group $\langle G,M,e\rangle$ to the group $\langle G',M',e'\rangle$ a group homomorphism?

Exercise 1.6. Suppose $h\colon \mathfrak{C} \to \mathfrak{D}$ is a homomorphism. Is the image of $\mathfrak{C}$ under h, $h(\mathfrak{C})$, a substructure of $\mathfrak{D}$?

Definition 1.9. The homomorphism $h\colon \mathfrak{A} \to \mathfrak{B}$ is an *isomorphism* if there is a homomorphism $h'\colon \mathfrak{B} \to \mathfrak{A}$ such that hh' is the identity map on $\mathfrak{B}$ and $h'h$ the identity on $\mathfrak{A}$. In this case we write $\mathfrak{A} \cong \mathfrak{B}$. h is an *embedding* if h is an isomorphism between $\mathfrak{A}$ and $h(\mathfrak{A})$.

Exercise 1.7. With the notation of definition 1.8 show that $h\colon \mathfrak{A} \to \mathfrak{B}$ is an isomorphism if and only if

(i) h is a bijection;

(ii) for each $i \in I$, $a_1,\ldots,a_{\lambda(i)} \in A$
$\langle a_1,\ldots,a_{\lambda(i)}\rangle \in R_i$ iff
$\langle h(a_1),\ldots,h(a_{\lambda(i)})\rangle \in S_i$;

(iii) for each $j \in J$, $a_1,\ldots,a_{\mu(j)} \in A$
$h(f_j(a_1,\ldots,a_{\mu(j)})) = g_j(h(a_1),\ldots,h(a_{\mu(j)}))$;

and *(iv)* $h(c_k) = d_k$ for each $k \in K$.

Exercise 1.8. Give an example of non-isomorphic structures $\mathfrak{A}$, $\mathfrak{B}$ of the same type such that there exists a homomorphism $h: \mathfrak{A} \to \mathfrak{B}$ with h a one-one map from A onto B.

1.2. THE FIRST ORDER LANGUAGE ASSOCIATED WITH A RELATIONAL STRUCTURE.

In the first section we formalized a general notion of 'relational (algebraic) structure' which was sufficient to embrace many types of structure studied by mathematicians. Propositions about particular structures, such as for example theorems concerning groups, are formulated by mathematicians in English (with some standard abbreviations). In order that such propositions may be formulated in a more precise and unambiguous way, a formalized language associated with a given structure is introduced.

Definition 1.10. The *first order language* $L(\mathfrak{A})$ for the structure $\mathfrak{A}\ (= \langle A, \{R_i\}_{i\in I}, \{f_j\}_{j\in J}, \{c_k\}_{k\in K}\rangle)$ consists of

(i) individual variables $v_0, v_1, \ldots, v_n, \ldots$ (for $n \in N$);

(ii) individual constants $\mathbf{c}_k$ for each $k \in K$;

(iii) a $\lambda(i)$-ary predicate letter $\mathbf{R}_i$ for each $i \in I$;

(iv) a $\mu(j)$-ary function symbol $\mathbf{f}_j$ for each $j \in J$;

(v) logical connectives $\neg$('not') and & ('and');

(vi) the universal quantifier $\forall$('for all');

and *(vii)* brackets (,).

We adopt the convention that a boldface symbol in the language corresponds to the light face symbol for the relation, function, or constant in the structure $\mathfrak{A}$ (e.g. the predicate $\mathbf{R}_i$ corresponds to the relation R_i). Notice that the language $L(\mathfrak{A})$ depends effectively only on the type of the structure $\mathfrak{A}$. The languages for two structures of the same (or similar) type are, up to alphabetic variants, identical. Thus we will sometimes write $L(\delta)$ for the language associated with all structures of type δ. L will denote an arbitrary first order language. We say that $\mathfrak{A}$ is a *realization* of the language $L(\mathfrak{A})$ (or $L(\delta)$ if $\mathfrak{A}$ is a structure of type δ).

Finite strings of symbols of L with particular well defined forms constitute the vocabulary of L. Terms, atomic formulae, and well formed formulae are inductively defined subsets of the set of all ordered finite subsets of symbols in L.

Definition 1.11. The set of terms of L, *Term* (L), is the smallest set X such that

(i) all individual variables $v_0, v_1, \ldots$ and individual constants, c_k for $k \in K$ are members of X

and *(ii)* if $t_1, \ldots, t_{\mu(j)} \in X$ then $f_j(t_1, \ldots, t_{\mu(j)}) \in X$ (for each $j \in J$).

Exercise 1.9. Verify that each term of L is a finite string of symbols of L.

Definition 1.12. The set of atomic formulae of L, *Atom* (L), consists of all finite strings of elements of L with the form $R_i(t_1, \ldots, t_{\lambda(i)})$ where $i \in I$ and $t_1, \ldots, t_{\lambda(i)} \in$ *Term* (L).

Definition 1.13. The set of well formed formulae of L, *Form* (L), is the smallest set Y such that

(i) all atomic formulae are members of Y

and *(ii)* if $\phi, \psi \in Y$ then $(\phi \;\&\; \psi), \neg\phi, \forall v_i \phi \in Y$.

Exercise 1.10. Verify that each well formed formula of L is a finite string of symbols of L.

As indicated in the definitions we use t (with or without subscripts) as a variable for a term of L and $\phi, \psi, \chi, \ldots$ as variables for well formed formulae. We sometimes use x as a variable ranging over the set of variables $\{v_i : i = 0, 1, 2, \ldots\}$. A 'well formed formula' is sometimes abbreviated to a 'formula'.

Definition 1.14. The set of subformulae of a formula ϕ, *Subform* (ϕ), is defined recursively:

$$Subform\ (R_i(t_1,\ldots,t_{\lambda(i)})) = \{R_i(t_1,\ldots\ t_{\lambda(i)})\};$$

$$Subform\ (\neg\phi) = \{\neg\phi\} \cup Subform\ (\phi);$$

$$Subform\ ((\phi\ \&\ \psi)) = \{(\phi\ \&\ \psi)\} \cup Subform\ (\phi) \cup Subform\ (\psi);$$

and

$$Subform\ (\forall v_i\phi) = \{\forall v_i\phi\} \cup Subform\ (\phi).$$

It is immediately apparent from the definition and exercise 1.10 that for any $\phi \in Form(L)$, *Subform* (ϕ) is a finite subset of $Form(L)$ and that ϕ is atomic if and only if *Subform* (ϕ) contains the single element ϕ.

The logical connectives $\vee$ ('or'), $\rightarrow$ ('implies'), $\leftrightarrow$ ('if and only if') and the quantifier $\exists$ ('there exists') are defined in terms of the primitives as follows:

$$(\phi \vee \psi) =_{Df} \neg(\neg\phi\ \&\ \neg\psi)$$

$$(\phi \rightarrow \psi) =_{Df} \neg(\phi\ \&\ \neg\psi)$$

$$(\phi \leftrightarrow \psi) =_{Df} ((\phi \rightarrow \psi)\ \&\ (\psi \rightarrow \phi))$$

$$\exists v_i\phi =_{Df} \neg\forall v_i\neg\phi$$

Note that $\vee$, $\rightarrow$, $\leftrightarrow$, $\exists$ are not elements of the language L but are simply used as convenient abbreviations in certain well-formed formulae.

Alternative formulations of a first order language L can be given using other sets of primitive connectives in place of $\{\neg, \&, \forall\}$. Each of the pairs $\{\neg, \rightarrow\}$, $\{\neg, \vee\}$ is a possible set of primitive propositional connectives; the remaining propositional connectives can be defined in terms of truth-functionally equivalent forms involving only the primitive connectives. Similarly, the existential quantifier $\exists$ may be taken as the primitive quantifier and the universal quantifier is then defined in terms of $\exists$ and $\neg$. The use of a minimal set of connectives is simply a convenient technical device which shortens some definitions and proofs, and the particular set chosen here is just one such minimal set.

A few more definitions related to the notion of a well formed formula ϕ are necessary.

Definition 1.15. The *scope* of (the exhibited occurrence of) the quantifier $\forall v_i$ in .... $\forall v_i\phi$ .... is ϕ.

Definition 1.16. An occurrence of a variable x in $\phi \in Form(L)$ is *bound* if and only if either it is immediately after the sign $\forall$ in ϕ or it is in the scope of a quantifier occurrence $\forall x$ in ϕ.
An occurrence of x which is not bound is *free*.

In the case where the occurrence of x is in the scope ψ of a quantifier $\forall x$ in ϕ, x is said to be bound by that quantifier occurrence provided x does not lie in the scope of a quantifier $\forall x$ in ψ (i.e. x is bound by the innermost quantifier occurrence within whose scope x lies). Thus for instance in the formula $\forall x \exists y \forall x R(x,y)$ the occurrence of x in $R(x,y)$ is bound by the quantifier $\forall x$ immediately preceding R and not by the leftmost quantifier $\forall x$. We illustrate the two preceding definitions with some examples.

Consider the formula: $\forall v_1 R_1(f(v_1),c)$ & $R_2(v_1,v_2,d))$. The scope of the quantifier $\forall v_1$ is $R_1(f(v_1),c)$. Hence the first two occurrences of the variable v_1 are bound whereas the third occurrence (in $R_2(v_1,v_2,d)$) is free. The single occurrence of v_2 is free.
In the following formula: $\exists v_2(R_1(v_2,c) \rightarrow R_2(v_1,f(v_1,v_2)))$ each occurrence of v_2 is bound whereas v_1 is free at each occurrence.

Bound occurrences of variables (in a first order formula) may be compared with 'dummy' variables that occur in other familiar mathematical expressions. For example, suppose $f\colon R \rightarrow R$ is a real-valued integrable function on the reals. Occurrences of x in the expression $f(x)$ are free whereas in the expression $\int_0^1 f(x)dx$ the occurrences of x in $f(x)$ become bound by $\int_0^1 \ldots dx$. The effect of the integral sign is similar to that of the quantifier; it binds the free occurrences of x occurring within its scope. Similarly in the expression $\sum_{n=1}^{\infty} a_n$ for an infinite convergent series the occurrences of n are bound, or dummy variables. In this case $\sum_{n=1}^{\infty}$ binds the variable n (which occurs free) in a_n.

The use of brackets (), in the definition of a formula (1.13) and of the defined connectives $\vee$, $\rightarrow$, $\leftrightarrow$ ensures that given any formula ϕ its subformulae are unique and can readily be extracted from ϕ. Consider the formula

$$(\neg(\phi \;\&\; \neg\psi) \;\&\; \chi).$$

Without brackets this becomes $\neg\phi \;\&\; \neg\psi \;\&\; \chi$ which is not a formula. We can insert brackets to give a formula but there is no unique way of doing this: besides the original formula $(\neg(\phi \;\&\; \neg\psi) \;\&\; \chi)$ we can have $((\neg\phi \;\&\; \neg\psi) \;\&\; \chi)$, $(\neg\phi \;\&\; (\neg\psi \;\&\; \chi))$, $(\neg\phi \;\&\; \neg(\psi \;\&\; \chi))$, $\neg((\phi \;\&\; \neg\psi) \;\&\; \chi)$ etc.. Sometimes it *is* possible to omit certain brackets in a formula without losing unique readability. We shall often leave off the outermost brackets of ϕ when the outermost connective is binary (i.e. &, $\vee$, $\rightarrow$, $\leftrightarrow$). If we adopt a standard convention other brackets may also be omitted. The propositional connectives fall into three groups: $\neg$, & and $\vee$, $\rightarrow$ and $\leftrightarrow$, each of which is considered more binding than the one succeeding it. This means that, for example,

$$((\neg\phi \;\&\; \psi) \rightarrow (\chi \vee \phi)) \quad \text{can be written}$$

$$\neg\phi \;\&\; \psi \;\rightarrow\; \chi \vee \phi$$

unambiguously, according to the convention. We cannot however drop any brackets in the formula

$$\neg(\neg\phi \;\&((\psi \rightarrow \chi) \vee \phi)).$$

Another standard convention is to write

$\phi_1 \;\&\; \ldots \;\&\; \phi_k$ or $\underset{1\le i\le k}{\&} \phi_i$ for $(\phi_1 \;\&(\phi_2 \;\&\ldots(\ldots(\&\; \phi_k)\ldots)$

and $\psi_1 \vee \ldots \vee \psi_k$ or $\bigvee_{1\le i\le k} \psi_i$ for $(\psi_1 \vee(\psi_2 \ldots(\ldots(\vee\; \psi_k)\ldots)$.

It is sometimes convenient to display for a particular $\phi \in$ *Form* (L) the variables which have free occurrences in ϕ. We will use the notation $\phi(x_1,\ldots,x_n)$ to denote that $\phi \in$ *Form* (L) and that the free variables in ϕ (necessarily a finite number since ϕ is of finite length) are included in the set $\{x_1,\ldots,x_n\}$. Similarly, if $t \in Term(L)$ we use $t(x_1,\ldots,x_n)$ to denote that the variables occurring in t are in $\{x_1,\ldots,x_n\}$. We will sometimes abbreviate this using vector notion: $\phi(\vec{x})$ stands for $\phi(x_1,\ldots,x_n)$ and $t(\vec{x})$ for $t(x_1,\ldots,x_n)$. Similarly a finite sequence of *like* quantifiers $\forall x_1 \ldots \forall x_n$ (or $\exists x_1 \ldots \exists x_n$) will sometimes be written $\forall\vec{x}$ (or $\exists\vec{x}$). If $\phi(\vec{x}) \in Form(L)$ then $\phi(\vec{t})$ in $Form(L)$ will denote the formula obtained from $\phi(\vec{x})$ when each free occurrence of x_i is replaced by t_i.

Consider the following formula: $\forall v_1\phi(v_1,v_2)$ (where v_2 has a free occurrence in ϕ). v_2 is an individual variable and indeed is meant to represent a variable in the sense that the substitution of another arbitrary variable for each free occurrence of the variable v_2 does not change the formula essentially. But suppose we substitute v_1 for v_2. Then the formula becomes $\forall v_1\phi(v_1,v_1)$ which no longer has a free variable. A similar situation arises if the term $f(v_1)$ is substituted for v_2. The following definition is designed to prevent precisely this kind of 'clash' occurring when an arbitrary term t is substituted for a free variable x in a formula ϕ.

<u>Definition 1.17.</u> *$t \in Term(L)$ is free for x in $\phi \in Form(L)$* iff the variable x has no free occurrences in ϕ which lie within the scope of a quantifier $\forall y$ where y is a variable occurring in t.

The definition ensures that if t is substituted for x in ϕ then no variable in t becomes bound, as it were, accidentally. In the example preceding the definition neither the term v_1 nor the term $f(v_1)$ is free for v_2 in $\forall v_1\phi(v_1,v_2)$. We remark that for any formula ϕ the variable x is free for

x in ϕ.

Exercise 1.11. In which, if any, of the following formulae is the term $f(v_1, v_2, c)$ free for v_2:

(i) $(\phi(v_1, v_2) \ \& \ \psi(v_2, c)) \rightarrow \forall v_1 \phi(v_1, c)$

(ii) $\phi(g(v_1, v_2)) \vee \forall v_1 \psi(g(v_1, c), v_1)$

(iii) $\forall v_1 \phi(v_1, v_2, v_3) \ \& \ \psi(v_1, v_2)$

(iv) $\forall v_1 \forall v_2 \phi(v_1, h(v_1, v_2))$?

Definition 1.18. If $\phi \in Form(L)$ and ϕ has no free variables then ϕ is a *sentence*. We use the abbreviation $Sent(L)$ for the subset of $Form(L)$ consisting of all sentences and σ as a variable for sentences.

Definition 1.19. If $\phi(x_1, \ldots, x_n) \in Form(L)$ then the *closure of* ϕ is $\forall x_1 \ldots \forall x_n \ \phi(x_1, \ldots, x_n) \in Sent(L)$.

We showed at the beginning of this section how to associate with a given relational structure $\mathfrak{A}$ (up to isomorphism) a unique language $L(\mathfrak{A})$. More precisely, there is a one-one correspondence between types of relational structures and first order languages. It is sometimes convenient to consider an extension L^* of a given first order language L formed by adding new non-logical symbols, predicate letters, function symbols, and individual constants. A realization $\mathfrak{A}$ for L will not in general be a realization for L^*. However $\mathfrak{A}$ can be expanded to a structure $\mathfrak{A}^*$ which will be a realization for L^*. The way this is accomplished is by the imposition of extra structure on the domain A of $\mathfrak{A}$. The structure $\mathfrak{A}^*$ will have the same domain as that of $\mathfrak{A}$. New relations and functions are then defined on A in one-one correspondence with the new predicate and function symbols in the language and similarly certain members of A are singled out as distinguished elements in correspondence with the new individual constants in the language. For example, consider the language

$L(\mathfrak{G})$ for a group $\mathfrak{G} = \langle G,=,+,0\rangle$. Any ring $\langle R,=,+,0\rangle$ presented as an additive group, is a realization for $L(\mathfrak{G})$. If a new function is added to $L(\mathfrak{G})$ to give L^*, the language for a ring, then the full structure of the ring $\langle R,=,+,0,\cdot\rangle$ is a realization for L^*. A structure $\mathfrak{A}^*$ obtained from the structure $\mathfrak{A}$ by the addition of new relations and functions on A and an enlargement of the set of distinguished elements is an *expansion* of $\mathfrak{A}$ and $\mathfrak{A}$ is a *reduct* of $\mathfrak{A}^*$.

1.3. INTERPRETATIONS.

A first order language L, even though associated with a given (type of) structure, is technically a syntactic object with no semantic significance. However the intention is to be able to express propositions concerning a given structure. It remains to be shown how this is to be achieved.

The variables are to be interpreted as ranging over the domain A. The constant symbols and function symbols are interpreted as the constants and functions with which they are in one-one correspondence. So the terms, relative to an assignment of elements of A to the variables, are interpreted as elements of A. The predicates are interpreted as the corresponding relations and the connectives have their intended meaning: '$\neg$' is 'not', '&' is 'and', and '$\forall x$' is 'for every element of A'. A given formula ϕ is interpreted, relative to an interpretation of the free variables, as an assertion about $\mathfrak{A}$. For example consider the cyclic group with six elements, $G = \{g,g^2,\ldots,g^6 = e\}$. Let $\mathfrak{G} = \langle G,=,\cdot,e\rangle$ and $\phi \in \mathrm{Form}(L(\mathfrak{G}))$ be the formula: $(v_1\cdot v_1)\cdot v_1 = e$. (We write $x\cdot y$ instead of $\cdot(x,y)$ for the binary function representing the group operation and $t_1=t_2$ for $=(t_1,t_2)$.) If v_1 is interpreted as e, g^2, or g^4 then ϕ makes a correct assertion about $\mathfrak{G}$ since e^3, $(g^2)^3$, $(g^4)^3$ are all equal to e. However if v_1 is interpreted as g, g^3, or g^5 then the assertion made by ϕ is no longer valid. The formula $\forall v_1((v_1\cdot v_1)\cdot v_1 = e)$ has no free variables and so makes an absolute assertion about $\mathfrak{G}$:

'for all $g \in G$, $g^3 = e$'

which is false.
Alternatively $\exists v_1((v_1 \cdot v_1) \cdot v_1 = \mathrm{e})$ makes a correct assertion because it is an abbreviation for

$$\neg \forall v_1 \neg ((v_1 \cdot v_1) \cdot v_1 = \mathrm{e}),$$

i.e. 'not for all $g \in G$ is it not the case that $g^3 = e$'
or 'there is some $g \in G$ such that $g^3 = e$'.

These informal ideas illustrated above on the intended meaning of the language $L(\mathfrak{A})$ are now incorporated in a formal definition. First let $\mathrm{a} = \langle a_0, a_1, \ldots \rangle$ be an infinite sequence of elements of the domain of the structure $\mathfrak{A}$. We do not insist that the elements a_i, as elements of A, be distinct and indeed each a_i may be the same element of A. The assignment determined by a is that in which the variable v_i is interpreted as the element $a_i \in A$. A further piece of notation we shall use is the following: $\mathrm{a}(b/_n)$ is an abbreviation for the sequence obtained from a with b replacing a_n, i.e. $\mathrm{a}(b/_n) = \langle a_0, a_1, \ldots, a_{n-1}, b, a_{n+1}, \ldots \rangle$.

First we show how to interpret terms of the language $L(\mathfrak{A})$ in the structure $\mathfrak{A}$ relative to the assignment a.

<u>Definition 1.20.</u> *The denotation of a term t in $\mathfrak{A}$ with respect to* a, $t^{\mathfrak{A}}[\mathrm{a}]$, (where a is the sequence $\langle a_0, a_1, \ldots \rangle$), is defined recursively as follows:

(i) if t is v_i then $t^{\mathfrak{A}}[\mathrm{a}] = a_i$;

(ii) if t is c_k then $t^{\mathfrak{A}}[\mathrm{a}] = c_k$;

(iii) if t is $\mathrm{f}_j(t_1, \ldots, t_{\mu(j)})$ then

$$t^{\mathfrak{A}}[\mathrm{a}] = f_j(t_1^{\mathfrak{A}}[\mathrm{a}], \ldots, t_{\mu(j)}{}^{\mathfrak{A}}[\mathrm{a}]).$$

This definition is very natural; each variable v_i is replaced by the element $a_i \in A$ and each constant and function symbol is replaced by the distinguished element or function to which it corresponds. Relative to a given assignment there is a unique element $t^{\mathfrak{A}}[\mathrm{a}] \in A$ associated with $t \in \mathrm{Term}(L(\mathfrak{A}))$.

Definition 1.21. a *satisfies* ϕ *in* $\mathfrak{A}$, $\mathfrak{A} \models_{a} \phi$, where $\phi \in Form(L(\mathfrak{A}))$, is defined recursively:

(i) $\mathfrak{A} \models_{a} R_i(t_1,\ldots,t_{\lambda(i)})$ iff $\langle t_1^{\mathfrak{A}}[a],\ldots,t_{\lambda(i)}^{\mathfrak{A}}[a]\rangle \in R_i$;

(ii) $\mathfrak{A} \models_{a} \neg\phi$ iff it is not the case that $\mathfrak{A} \models_{a} \phi$;

(iii) $\mathfrak{A} \models_{a} \phi_1 \& \phi_2$ iff $\mathfrak{A} \models_{a} \phi_1$ and $\mathfrak{A} \models_{a} \phi_2$;

(iv) $\mathfrak{A} \models_{a} \forall v_i\phi$ iff for any $b \in A$ $\mathfrak{A} \models_{a(b/_i)} \phi$.†

Definitions 1.20 and 1.21 may be thought of as providing a dictionary which, in a given context, 'translates' the formal language $L(\mathfrak{A})$ into English. The logical connectives, which are independent of the context, $\neg$, & are given their intended meanings, 'not', 'and'. As a corollary to the definition the defined connectives $\rightarrow$, $\vee$, $\leftrightarrow$ also have their intended interpretation under this definition. Likewise the existential quantifier $\exists$ is interpreted as 'there exists an element of A such that....' just as (1.21 *(iv)*) $\forall$ is interpreted as 'for every element of A....'.

Lemma 1.22. *For every* ϕ, $\psi \in Form(L(\mathfrak{A}))$ *and every sequence* a *in* A

(a) $\mathfrak{A} \models_{a} \phi \vee \psi$ *iff* $\mathfrak{A} \models_{a} \phi$ *or* $\mathfrak{A} \models_{a} \psi$;

(b) $\mathfrak{A} \models_{a} \phi \rightarrow \psi$ *iff when* $\mathfrak{A} \models_{a} \phi$ *then* $\mathfrak{A} \models_{a} \psi$ *iff not* $\mathfrak{A} \models_{a} \phi$ *or* $\mathfrak{A} \models_{a} \psi$;

(c) $\mathfrak{A} \models_{a} \phi \leftrightarrow \psi$ *iff* $\mathfrak{A} \models_{a} \phi$ *just in case* $\mathfrak{A} \models_{a} \psi$;

(d) $\mathfrak{A} \models_{a} \exists v_i\phi$ *iff for some* $b \in A$ $\mathfrak{A} \models_{a(b/_i)} \phi$.

Proof: *(a)* $\phi \vee \psi$ is by definition $\neg(\neg\phi \& \neg\psi)$. So by 1.21

---

†The definition of satisfaction using an infinite sequence a of the domain is due to Tarski (1935). Alternative forms of definitions 1.20 and 1.21 can be given using only a finite sequence of elements of the domain as in e.g. Chang and Keisler (1973).

$\mathfrak{A} \models_a \phi \vee \psi$ iff it is not the case that $\mathfrak{A} \models_a \neg\phi \,\&\, \neg\psi$

iff it is not the case that $\mathfrak{A} \models_a \neg\phi$ and $\mathfrak{A} \models_a \neg\psi$.

By 1.21 *(ii)* it is not the case that $\mathfrak{A} \models_a \neg\phi$ just when $\mathfrak{A} \models_a \phi$. So $\mathfrak{A} \models_a \phi \vee \psi$ iff either $\mathfrak{A} \models_a \phi$ or $\mathfrak{A} \models_a \psi$.

(b) and *(c)* are left as exercises.

To prove *(d)* we first remark that $\mathfrak{A}\ \exists v_i \phi$ is by definition $\neg \forall v_i \neg \phi$. Then definition 1.21 implies that

$\mathfrak{A} \models_a \exists v_i \phi$ iff it is not the case that $\mathfrak{A} \models_a \forall v_i \neg \phi$

iff it is not the case that for all $b \in A$ $\mathfrak{A} \models_{a(b/_i)} \neg\phi$

iff for some $b \in A$ it is not the case that $\mathfrak{A} \models_{a(b/_i)} \neg\phi$

iff for some $b \in A$ $\mathfrak{A} \models_{a(b/_i)} \phi$. □

To illustrate the formal definition of satisfaction consider the structure $\langle Q,<\rangle$ with domain Q, the rationals, ordered by the relation $<$. Then $L(\mathfrak{Q})$ contains a single binary relation $<$ corresponding to the relation $<$.

Suppose ϕ is $\exists v_1(v_0<v_1 \,\&\, v_1<v_2 \,\&\, \forall v_3(v_1<v_3 \rightarrow v_0<v_3)$. We write $\mathfrak{Q}$ for $\langle Q,<\rangle$ and q for a sequence of rationals. By definition 1.21 and lemma 1.22

$\mathfrak{Q} \models_q \phi$ iff there is a rational x such that

$\mathfrak{Q} \models_{q(x/_1)} v_0<v_1 \,\&\, v_1<v_2 \,\&\, \forall v_3(v_1<v_3 \rightarrow v_0<v_3)$

iff there is a rational x such that

(i) $\mathfrak{Q} \models_{q'} v_0<v_1$,

(ii) $\mathfrak{Q} \models_{q'} v_1<v_2$,

and *(iii)* for every rational y if

$\mathfrak{Q} \models_{q'(y/_3)} v_1<v_3$ then $\mathfrak{Q} \models_{q'(y/_3)} v_0<v_3$

(where q' is $q(x/_1)$)

iff there is a rational x such that

(i) $v_0^{\mathfrak{Q}}[q'] < v_1^{\mathfrak{Q}}[q'] < v_2^{\mathfrak{Q}}[q']$

and *(ii)* for every rational y if

$$v_1^{\mathfrak{Q}}[\mathbf{q}'(y/_3)] < v_3^{\mathfrak{Q}}[\mathbf{q}'(y/_3)]$$
$$\text{then } v_0^{\mathfrak{Q}}[\mathbf{q}'(y/_3)] < v_3^{\mathfrak{Q}}[\mathbf{q}'(y/_3)].$$

Now we use definition 1.20:

$$v_i^{\mathfrak{Q}}[\mathbf{q}'] = q_i' = \begin{cases} q_0 & \text{if } i = 0; \\ x & \text{if } i = 1 \\ q_2 & \text{if } i = 2 \end{cases} \quad v_j^{\mathfrak{Q}}[\mathbf{q}'(y/_3)] = \begin{cases} q_0 & \text{if } j = 0 \\ x & \text{if } j = 1 \\ y & \text{if } j = 3 \end{cases}$$

Combining these results we have:

$\mathfrak{Q} \models_{\mathbf{q}} \phi$ iff there is a rational x such that

(i) $q_0 < x < q_2$

and *(ii)* for every rational y if $x < y$ then $q_0 < y$.

Now if $q_0 < x$ then necessarily for every rational $y, x < y$ implies $q_0 < y$. So the condition on the right above will be satisfied just in case there is a rational x with $q_0 < x < q_2$. So finally, $\mathfrak{Q} \models_{\mathbf{q}} \phi$ iff $q_0 < q_2$.

Exercise 1.12. For which sequences of rationals $\mathbf{q}$ does $\langle Q,<\rangle \models_{\mathbf{q}} \phi$ where ϕ is *(a)* $v_0<v_1 \;\&\; \exists v_2(v_1<v_2)$

(b) $\forall v_0(v_0<v_1 \;\&\; v_1<v_2 \rightarrow v_0<v_2)$?

Exercise 1.13. For how many groups $\langle G,=,\cdot,e\rangle$ (up to isomorphism) does

$$\langle G,=,\cdot,e\rangle \models_{\mathbf{g}} v_0\cdot v_1 = v_1\cdot v_0 \;\&\; \exists v_0 \exists v_1 \exists v_2 \exists v_3 \forall v_4 (v_4 = v_0 \vee v_4 = v_1 \vee v_4 = v_2 \vee v_4 = v_3)$$

for all sequences $\mathbf{g}$?

It can be seen from the example $\langle Q,<\rangle \models_{\mathbf{q}} \phi$ where ϕ is $\exists v_1(v_0<v_1 \;\&\; v_1<v_2 \;\&\; \forall v_3(v_1<v_3 \rightarrow v_0<v_3)$ that the only elements of the sequence $\mathbf{q}$ that affect the truth value of $\langle Q,<\rangle \models_{\mathbf{q}} \phi$ are q_0 and q_2, the elements corresponding to the free variables v_0 and v_2 in ϕ. An inspection of definitions 1.20 and 1.21 shows immediately that the only elements of the sequence $\mathbf{a}$ that affect the truth value of $\mathfrak{A} \models_{\mathbf{a}} \phi$ are those a_i such that v_i occurs in ϕ. The following theorem shows that in fact the only elements a_i which affect the truth value are those such that v_i is *free* in ϕ.

Theorem 1.23. *Suppose* $v_{i_o},\ldots,v_{i_k}$ *are the free variables of* ϕ.
Then for any a, $\mathfrak{A} \models_{\mathrm{a}} \phi$ *iff for all sequences* b

$$\mathfrak{A} \models_{\mathrm{b}(a_{i_o}/i_o,\ldots,a_{i_k}/i_k)} \phi.$$

Proof: We first remark that a consequence of definition 1.20 is that $t^{\mathfrak{A}}[\mathrm{a}] = t^{\mathfrak{A}}[\mathrm{b}]$ if $a_i=b_i$ for all i such that v_i occurs in the term t. We prove the theorem by induction on the length of ϕ (which may be measured by the number of connectives and quantifiers in ϕ). Note that a subformula ϕ' of the formula ϕ may have free variables which are not free in ϕ.

First we consider the case where ϕ is atomic (i.e. has length zero). Then ϕ is of the form $\mathrm{R}_i(t_1,\ldots,t_{\lambda(i)})$. The free variables of ϕ are those variables occurring in $t_1,\ldots,t_{\lambda(i)}$. Suppose these are $v_{i_o},\ldots,v_{i_k}$. Then by the preliminary remark, for any sequences a ,b

$$t_r^{\mathfrak{A}}[\mathrm{a}] = t_r^{\mathfrak{A}}[\mathrm{b}(a_{i_o}/i_o,\ldots,a_{i_k}/i_k)] \text{ for } r=1,\ldots,\lambda(i).$$

Definition 1.21*(i)* then implies that for any sequences a ,b $\mathfrak{A} \models_{\mathrm{a}} \mathrm{R}_i(t_1,\ldots,t_{\lambda(i)})$ iff

$$\mathfrak{A} \models_{\mathrm{b}(a_{i_o}/i_o,\ldots,a_{i_k}/i_k)} \mathrm{R}_i(t_1,\ldots,t_{\lambda(i)}).$$

Now suppose that ϕ is $\neg\psi$. The free variables of ϕ are those of ψ and by the induction hypothesis, since ψ has fewer connectives and quantifiers than ϕ, for any a

$\mathfrak{A} \models_{\mathrm{a}} \psi$ iff for all sequences b

$$\mathfrak{A} \models_{\mathrm{b}(a_{i_o}/i_o,\ldots,a_{i_k}/i_k)} \psi.$$

For any formula χ either $\mathfrak{A} \models_{\mathrm{a}} \chi$ or $\mathfrak{A} \models_{\mathrm{a}} \neg\chi$ and so $\mathfrak{A} \models_{\mathrm{a}} \phi$ iff for all sequences **b**, $\mathfrak{A} \models_{\mathrm{b}(a_{i_o}/i_o,\ldots,a_{i_k}/i_k)} \phi$.

The case where ϕ is ϕ_1 & ϕ_2 is equally straightforward, using the induction hypothesis for the formulae ϕ_1, ϕ_2, and will be omitted.

Lastly suppose that ϕ is $\forall v_j \psi$. Then v_j is not a free variable of ϕ but may be a free variable of ψ. By definition $\mathfrak{A} \models_{\mathrm{a}} \phi$ iff $\mathfrak{A} \models_{\mathrm{a}(b/j)} \psi$ for all $b \in A$. The induction hypothesis applied to ψ yields

$\mathfrak{A} \models_{\mathrm{a}(b/j)} \psi$ iff for all sequences b

$$\mathfrak{A} \models_{\mathrm{b}(b/j, a_{i_o}/i_o, \ldots, a_{i_k}/i_k)} \psi.$$

So $\mathfrak{A} \models_{\mathrm{a}} \phi$ iff for all $b \in A$ and all sequences b

$$\mathfrak{A} \models_{\mathrm{b}(b/j, a_{i_o}/i_o, \ldots, a_{i_k}/i_k)} \psi$$

and hence $\mathfrak{A} \models_{\mathrm{a}} \phi$ iff for all sequences b

$$\mathfrak{A} \models_{\mathrm{b}(a_{i_o}/i_o, \ldots, a_{i_k}/i_k)} \phi. \quad \square$$

The method of proof used in theorem 1.23, induction on the length of a formula, illustrates a fundamental principle that is frequently used to establish properties of all formulae of L. There are four cases to consider corresponding to the clauses of definition 1.13. First the property is established for atomic formulae and then, assuming the property to hold for all proper subformulae of ϕ, it is established for ϕ first of the form $\neg\psi$, then of the form $\phi_1 \& \phi_2$, and finally of the form $\forall v_i \psi$. The complexity of a formula ϕ can be measured by the number of connectives and quantifiers in ϕ and induction on this natural number is used to establish properties of ϕ.

<u>Corollary 1.24.</u> *For all* $\sigma \in Sent(L(\mathfrak{A}))$

either $\mathfrak{A} \models_{\mathrm{a}} \sigma$ *for all sequences* a
or $\mathfrak{A} \models_{\mathrm{a}} \neg\sigma$ *for all sequences* a..

Proof: A sentence σ by definition has no free variables and so this corollary follows immediately from theorem 1.23. $\square$

Following the definition of free and bound variables (1.16) certain analogies were noted between occurrences of bound variables in first order formulae and dummy variables used elsewhere in mathematics. The first example cited there was the expression for the definite integral $\int_0^1 f(x)dx$. This integral may equally well be represented by the expression

$\int_0^1 f(y)dy$ (provided y does not occur in the expression $f(x)$). The switch of dummy variable from x to the new variable y does not alter the meaning of the expression. Similarly the two infinite sums $\sum_{n=1}^{\infty} a_n$ and $\sum_{m=1}^{\infty} a_m$ (where m does not occur in the expression a_n) represent the same real number. A precisely similar situation occurs when bound occurrences of a variable v_i in a formula ϕ are replaced by another variable v_j not occurring in ϕ. The meaning of ϕ in any given interpretation is unchanged.

This is established in the following lemma.

__Lemma 1.25.__ *Suppose v_j does not occur in ϕ and that ϕ^i_j is the formula resulting when each bound occurrence in ϕ of v_i is replaced by v_j. Then for any structure $\mathfrak{A}$ and sequence* a

$$\mathfrak{A} \models_{\mathrm{a}} \phi \quad \textit{iff} \quad \mathfrak{A} \models_{\mathrm{a}} \phi^i_j$$

Proof: Again we use induction on the length of ϕ. In the case where ϕ is atomic there is nothing to prove for then ϕ does not contain bound occurrences of any variable and so ϕ^i_j is the same formula as ϕ. When ϕ is $\neg\psi$ then ϕ^i_j is $\neg\psi^i_j$. In this case the result follows immediately from the induction hypothesis and the satisfaction definition. The case where ϕ is a conjunction ϕ_1 & ϕ_2 is equally straightforward. Now ϕ^i_j is $(\phi_1)^i_j$ & $(\phi_2)^i_j$ and the induction hypothesis gives us $\mathfrak{A} \models_{\mathrm{a}} \phi_k$ iff $\mathfrak{A} \models_{\mathrm{a}} (\phi_k)^i_j$ for $k=1,2$. Then we apply definition 1.21 to infer the result $\mathfrak{A} \models_{\mathrm{a}} \phi$ iff $\mathfrak{A} \models_{\mathrm{a}} \phi^i_j$. Finally we consider the tricky case when ϕ is $\forall v_p \psi(v_p)$. We must distinguish between the cases when p is i and when p is not i. The second case is simpler for then ϕ^i_j is $\forall v_p \psi^i_j(v_p)$. The induction hypothesis implies that $\mathfrak{A} \models_{\mathrm{a}(b/_p)} \psi(v_p)$ iff $\mathfrak{A} \models_{\mathrm{a}(b/_p)} \psi^i_j(v_p)$ (where $\mathrm{a}(b/_p)$ is an arbitrary sequence of elements in the domain of $\mathfrak{A}$). Hence

$\mathfrak{A} \models_{\mathrm{a}(b/_p)} \psi(v_p)$ for all $b \in A$ just in case

$\mathfrak{A} \models_{\mathrm{a}(b/_p)} \psi^i_j(p)$ for all $b \in A$. But then by 1.21 we have

$\mathfrak{A} \models_a \phi$ iff $\mathfrak{A} \models_a \phi^i_j$. Lastly we consider the case where ϕ is $\forall v_i \psi(v_i)$. Now ϕ^i_j is $\forall v_j \psi^i_j(v_j)$. By the induction hypothesis $\mathfrak{A} \models_a \psi(v_i)$ iff $\mathfrak{A} \models_a \psi^i_j(v_i)$. Since v_j does not occur in ϕ it also has no occurrences in $\psi(v_i)$. Now a simple inductive proof shows that

$\mathfrak{A} \models_a \psi(v_i)$ iff $\mathfrak{A} \models_{a(a_i/_j, a_j/_i)} \psi(v_j)$ Hence for all sequences a

$\mathfrak{A} \models_{a(b/_j, a_j/_i)} \psi(v_j)$ iff $\mathfrak{A} \models_{a(b/_j, a_j/_i)} \psi^i_j(v_j)$.

Now again using the satisfaction definition $\mathfrak{A} \models_a \phi$ iff $\mathfrak{A} \models_a \phi^i_j$. This completes the proof. □

Exercise 1.14. Construct a formula ϕ with free occurrences of v_j such that for some structure $\mathfrak{A}$ and sequence a

$$\mathfrak{A} \models_a \phi \text{ and not } \mathfrak{A} \models_a \phi^i_j.$$

Lemma 1.26. *Suppose $t \in Term(L)$ and t is free for v_i in ϕ. Then $\mathfrak{A} \models_a \phi(t)$ iff $\mathfrak{A} \models_{a(b/_i)} \phi(v_i)$ where $t^{\mathfrak{A}}[a] = b$.*

Proof: Suppose the variables occurring in t are $v_{i_1}, \ldots, v_{i_k}$. Since t is free for v_i in ϕ there are no free occurrences of v_i in ϕ lying within the scope of a quantifier $\forall v_{i_j}$ $(j=1,\ldots,k)$. However there may be bound occurrences of the variables v_{i_j} in ϕ. If a straightforward induction on the length of ϕ is used to establish the lemma such bound occurrences are tiresome. The proof can be simplified by eliminating those occurrences at the outset. Accordingly we define $\phi^*(v_i)$ to be the formula which results when each bound occurrence of v_{i_j} in $\phi(v_i)$ is replaced by a new variable v_{r_j} not already occurring in $\phi(v_i)$ or t. A consequence of the hypothesis that t be free for v_i in ϕ is that $\phi^*(t)$ is the formula obtained when each bound occurrence of v_{i_j} in $\phi(t)$ is replaced by v_{r_j} for no occurrences of v_{i_j} become bound as it were accidentally when t is put in place of free occurrences of v_i. A corollary of the preceding lemma is that

$$\mathfrak{A} \models_{a(b/_i)} \phi(v_i) \quad \text{iff} \quad \mathfrak{A} \models_{a(b/_i)} \phi^*(v_i)$$

and

$$\mathfrak{A} \models_a \phi(t) \quad \text{iff} \quad \mathfrak{A} \models_a \phi^*(t).$$

Hence it will suffice to show that

$$\mathfrak{A} \models_a \phi^*(t) \quad \text{iff} \quad \mathfrak{A} \models_{a(b/_i)} \phi^*(v_i) \quad \text{where} \quad t^{\mathfrak{A}}[a] = b.$$

We use induction on the length of ψ to show that for all subformulae ψ of $\phi^*(v_i)$

$$\mathfrak{A} \models_a \psi(t) \quad \text{iff} \quad \mathfrak{A} \models_{a(b/_i)} \psi(v_i) \quad \text{where} \quad t^{\mathfrak{A}}[a] = b.$$

In the case where ψ is atomic the result is an immediate consequence of definition 1.21*(i)*. If ψ is of the form $\neg\psi'$ or $\psi_1 \;\&\; \psi_2$ the property follows directly using the induction hypothesis and the satisfaction definition. When ψ is of the form $\forall v_p \psi_1(v_p)$ we know that v_p is not one of the variables of t (since v_p will be bound in $\phi^*(v_i)$). The induction hypothesis implies that

$$\mathfrak{A} \models_a \psi_1(v_p, t) \quad \text{iff} \quad \mathfrak{A} \models_{a(b/_i)} \psi_1(v_p, v_i) \quad \text{where} \quad t^{\mathfrak{A}}[a] = b$$

for all sequences a.

Since v_p does not occur in t we have

$$t^{\mathfrak{A}}[a] = t^{\mathfrak{A}}[a(c/_p)] \quad \text{for all} \quad c \in A.$$

(Intuitively, the interpretation of t is independent of that of v_p.) So for all $c \in A$

$$\mathfrak{A} \models_{a(c/_p)} \psi_1(v_p, t) \quad \text{iff} \quad \mathfrak{A} \models_{a(c/_p, b/_i)} \psi_1(v_p, v_i)$$

where $t^{\mathfrak{A}}[a] = b$.

So by the satisfaction definition

$$\mathfrak{A} \models_a \psi(t) \quad \text{iff} \quad \mathfrak{A} \models_{a(b/_i)} \psi(v_i) \quad \text{for} \quad b = t^{\mathfrak{A}}[a].$$

This completes the proof. □

Theorem 1.23 justifies the following abbreviations in notation:

(i) If $t(v_{i_o}, \ldots, v_{i_k}) \in \mathit{TermL}(\mathfrak{A})$ we write

$t^{\mathfrak{A}}[a_{i_o},\ldots,a_{i_n}]$ for $t^{\mathfrak{A}}[\mathrm{a}]$.

(ii) If $\sigma \in Sent(L(\mathfrak{A}))$ and $\mathfrak{A} \models_{\mathrm{a}} \sigma$ for some sequence a then, since $\mathfrak{A} \models_{\mathrm{a}} \sigma$ for all sequences a, we write $\mathfrak{A} \models \sigma$.

(iii) If $\phi(v_{i_o},\ldots,v_{i_k}) \in Form(L(\mathfrak{A}))$ and $\mathfrak{A} \models_{\mathrm{a}} \phi(v_{i_o},\ldots,v_{i_k})$ then we write $\mathfrak{A} \models \phi[a_{i_o},\ldots,a_{i_k}]$.

Notice that $\phi[a_{i_o},\ldots,a_{i_k}]$ is *not* a formula of the language. It is shorthand for 'interpret the variable v_j as a_j'. The square brackets are not part of the language $L(\mathfrak{A})$ so even if $a_{i_o},\ldots,a_{i_k}$ happen to be constants of the language it is not the case that $\phi[a_{i_o},\ldots,a_{i_k}] \in Form\ L(\mathfrak{A})$.

Using vector notation we shall sometimes write $\vec{a} \in A$ and $\mathfrak{A} \models \phi[\vec{a}]$ to denote that $\phi(\vec{x})$ is satisfied in $\mathfrak{A}$ with the element $a_i \in A$ assigned to the variable x_i.

Exercise 1.15. Suppose $\mathfrak{A}, \mathfrak{B}$ are structures of the same type and $h : \mathfrak{A} \to \mathfrak{B}$ is an isomorphism. If $\phi(x_1,\ldots,x_k) \in Form\ L(\mathfrak{A})$ and $a_1,\ldots,a_k \in A$ show that

$$\mathfrak{A} \models \phi[a_1,\ldots,a_k] \quad \text{iff} \quad \mathfrak{B} \models \phi[h(a_1),\ldots,h(a_k)].$$

Exercise 1.16. *(a)* Suppose $\phi(x) \in Form\ L(\mathfrak{Q})$ where Q is the set of rationals and $\mathfrak{Q} = \langle Q,<\rangle$. Show that if $\mathfrak{Q} \models \phi[q]$ for some $q \in Q$ then $\mathfrak{Q} \models \phi[q']$ for any $q' \in Q$.

(b) Suppose $\phi(x_1,\ldots,x_h) \in Form\ L(\mathfrak{Q})$ with $\mathfrak{Q}$ as in *(a)*. Show that if $\mathfrak{Q} \models \phi[q_1,\ldots,q_h]$ and for i, $j=1,\ldots,h$, $q_i' < q_j'$ just in case $q_i < q_j$ then $\mathfrak{Q} \models \phi[q_1',\ldots,q_h']$.

The following definitions, which depend on definitions 1.20 and 1.21 characterize some formulae of L according to their semantic properties. We suppose that $\mathfrak{A}, \mathfrak{B}$ are realizations for L.

Definition 1.27.

(a) $\phi \in Form(L)$ is *valid* (or *true*) in $\mathfrak{A}$ just in case

$\mathfrak{A} \models_{a} \phi$ for all sequences a. When ϕ is valid in $\mathfrak{A}$ we say $\mathfrak{A}$ is a *model* for ϕ.

Note that if $\sigma \in Sent(L)$ then either σ or $\neg\sigma$ (but not both) is valid in $\mathfrak{A}$.

(b) $\phi \in Form(L)$ is *universally* (or *logically*) *valid* if it is valid in all realizations for L.

(c) $\phi \in Form(L)$ is *satisfiable* if for some realization $\mathfrak{B}$ for L and some sequence b in B $\mathfrak{B} \models_{b} \phi$.

(d) $\phi \in Form(L)$ is *refutable* if $\neg\phi$ is satisfiable.

(e) $\phi \in Form(L)$ is *contravalid* if $\neg\phi$ is universally valid.

Exercise 1.17. Suppose $\phi \in Form(L)$.

(a) Show that ϕ is universally valid if and only if $\neg\phi$ is not satisfiable.

(b) Show that $\phi(x_1,\ldots,x_n)$ is valid in $\mathfrak{B}$ if and only if $\forall x_1 \ldots . \forall x_n \phi(x_1,\ldots,x_n)$ is valid in $\mathfrak{B}$ and that $\phi(x_1,\ldots,x_n)$ is satisfiable in $\mathfrak{B}$ if and only if $\exists x_1 \ldots . \exists x_n \phi(x_1,\ldots,x_n)$ is satisfiable in $\mathfrak{B}$.

The rigorous definition of a true proposition in a structure $\mathfrak{A}$ is intended to capture the informal notion of truth as used by mathematicians. Further, a universally valid formula is one that is true on account of its logical form alone irrespective of the meaning of the predicates, functions, and constants in ϕ. It is the counterpart, in a first order language, of a tautology in a propositional language. There are many similarities between the two notions, tautologies and universally valid formulae, but there is one very important difference. Given a formula ϕ in a propositional language L_o (without quantifiers) there is a simple procedure to determine whether ϕ is a tautology. It is sufficient to examine the truth table of ϕ; ϕ is a tautology if and only if every assignment of truth values makes ϕ true. In general however there is no finite procedure to determine of a given first order formula whether or not it is universally valid. There are infinitely many non-isomorphic realizations of a given first order language because the domain may be of arbitrary cardinality. This contrasts with the finite number of distinct assignments of truth values to the variables in a formula in a propositional language. Although it is sometimes possible to establish that a given formula is not valid in some interpretation (and hence that it is not universally valid) it is not in general possible to ascertain in a finite way that a given formula is valid in every interpretation. Although the most obvious approach for determining the validity of a given formula in every situation (by considering every realization of the language of the formula) clearly fails there could, in principle, be other methods for establishing universal validity. However, a theorem due to Church (1936) proves that the problem is essentially undecidable, in the sense that no such general method exists. It is immediate from the definitions that a formula ϕ is not satisfiable just in case the formula $\neg\phi$ is universally valid and so in general the question of the satisfiability of a given formula is also undecidable.

The following lemmas illustrate some cases where the validity of a given formula is decidable.

Lemma 1.28. *Given a formula ϕ there are only finitely many non-isomorphic realizations of the language of ϕ with a domain of (finite) cardinality N.*

Proof: Suppose $R_1,\ldots,R_m$ are the distinct predicates occurring in ϕ, $f_1,\ldots,f_n$ the function symbols, and $c_1,\ldots,c_p$ the constants. Suppose R_i is a $\lambda(i)$-ary predicate. There are $N^{\lambda(i)}$ $\lambda(i)$-tuples from an N-element domain and hence $2^{(N^{\lambda(i)})}$ distinct subsets of $\lambda(i)$-tuples from an N-element domain; so there are $2^{(N^{\lambda(i)})}$ possible interpretations for R_i. Similarly the domain of the interpretation of the $\mu(j)$-ary function symbol f_j has $N^{\mu(j)}$ elements, the totality of distinct $\mu(j)$-tuples over an N-element domain. There are N possible values for each argument and hence $N^{(N^{\mu(j)})}$ possible interpretations of the function symbol f_j. There are N possible interpretations for each constant symbol c_i. Thus there are (allowing for automorphisms) at most

$$\frac{\prod_{i=1}^{m} 2^{(N^{\lambda(i)})} \times \prod_{j=1}^{n} N^{(N^{\mu(j)})} \times N^{p}}{N!}$$ possible

interpretations for the language of ϕ. □

The validity of a formula ϕ with a particular form can sometimes be determined by its validity in all domains of bounded finite cardinality. By lemma 1.28 there are only finitely many such (non-isomorphic) realizations for the language of ϕ. Further, for any such realization, $\mathfrak{A}$ say, with finite domain $\{s_1,\ldots,s_k\}$ there is a decision procedure to determine whether or not $\mathfrak{A} \models_{\mathrm{a}} \phi$. In this case definition 1.21 (iv) becomes:

$$\mathfrak{A} \models_{\mathrm{a}} \forall v_i \psi \quad \text{iff} \quad \mathfrak{A} \models_{\mathrm{a}(s_1/i)} \psi \text{ and....and } \mathfrak{A} \models_{\mathrm{a}(s_k/i)} \psi$$

from which it is clear that $\mathfrak{A} \models_{\mathrm{a}} \phi$ may be determined in a finite number of steps. Thus the problem of whether such a formula ϕ is universally valid is decidable.

<u>Lemma 1.29.</u> *Suppose* $\phi \in Form(L)$ *contains only unary predicate letters* $R_1,\ldots,R_k$ *and no functions or constants. Then* ϕ *is universally valid if and only if* ϕ *is valid in all domains with* $\leq 2^k$ *elements.*

Proof: Suppose $\mathfrak{A} = \langle A, R_1,\ldots,R_k\rangle$ is a realization for the language of ϕ such that for some $a_1,\ldots,a_n \in A$ $\mathfrak{A} \models \neg\phi[a_1,\ldots,a_n]$. We show that there is a realization $\mathfrak{A}^*$ with domain of at most 2^k elements in which ϕ is not valid. First we define an equivalence relation on A:

$$a \sim a' \text{ iff for each } i=1,\ldots,k \quad a \in R_i \text{ just in case } a' \in R_i.$$

It is clear that $\sim$ is an equivalence relation on A. Let A^* be the set of equivalence classes $\{[a]\}$ of A induced by $\sim$.

There are at most 2^k possible equivalence classes since for each $i=1,\ldots,k$ and any a either $a \in R_i$ or $a \notin R_i$. For $i=1,\ldots,k$ define $R_i^* = \{[a] : a \in R_i\}$. This is well defined since $[a] = [b]$ implies that a and b belong to precisely the same subsets R_i of A. (We call such a relation $\sim$ a congruence relation with respect to $R_1,\ldots,R_k$). Then $\mathfrak{A}^* = \langle A^*, R_1^*,\ldots,R_k^*\rangle$ is a realization for the language of ϕ. A straightforward induction on the length of a formula $\psi(x_1,\ldots,x_m)$ then shows that for any $a_1,\ldots,a_m \in A$

$$\mathfrak{A} \models \psi[a_1,\ldots,a_m] \quad \text{iff} \quad \mathfrak{A}^* \models \psi[[a_1],\ldots,[a_m]].$$

So then $\mathfrak{A}^* \models \neg\phi[[a_1],\ldots,[a_n]]$. Since the domain of $\mathfrak{A}^*$ has at most 2^k elements we have established that if ϕ is not valid in some structure then it is not valid in some structure with at most 2^k elements. The converse implication is trivial. □

We now state a very simple lemma which will be used in establishing the decidability of another class of formulae.

<u>Lemma 1.30.</u> *Suppose that* $\phi(x_1,\ldots,x_n) \in Form(L(\mathfrak{A}))$ *where* $\phi(x_1,\ldots,x_n)$ *is quantifier-free and that* $\mathfrak{A}^* \subseteq \mathfrak{A}$. *Then for any* $a_1,\ldots,a_n \in A^*$

$$\mathfrak{A} \models \phi[a_1,\ldots,a_n] \quad \text{iff} \quad \mathfrak{A}^* \models \phi[a_1,\ldots,a_n].$$

Proof: By hypothesis $\mathfrak{A}^*$ is a substructure of $\mathfrak{A}$ and hence $a_1,\ldots,a_n \in A \cap A^*$. An induction on the length of $t \in Term(L(\mathfrak{A}))$ shows that $t^{\mathfrak{A}}[a_1,\ldots,a_n] = t^{\mathfrak{A}^*}[a_1,\ldots,a_n]$ (when the variables in t are among $x_1,\ldots,x_n$). A second induction on the length of the (quantifier-free) formula ϕ is used to prove the lemma. The atomic case is an immediate consequence of our preliminary result on terms. For ϕ of the form $\neg\psi$ or $\phi_1 \,\&\, \phi_2$ the result is a simple consequence of definition 1.21 and the induction hypothesis. □

Theorem 1.31. *Suppose* $\sigma \in Sent(L)$ *is of the form* $\exists x_1 \cdots \exists x_n \forall \vec{y}\psi$ *where* ψ *contains no quantifiers, function symbols or constants. If* σ *is satisfiable then* σ *is satisfiable in a domain with at most n elements.*

Proof: Suppose $R_1,\ldots,R_k$ are the predicate letters occurring in ψ and that $\mathfrak{A} = \langle A,R_1,\ldots,R_k\rangle$ is a realization such that $\mathfrak{A} \models \sigma$. Then for some $a_1,\ldots,a_n \in A$ $\mathfrak{A} \models \forall\vec{y}\psi[a_1,\ldots,a_n]$. But then

$$(*) \quad \text{for all } \vec{b} \in A \quad \mathfrak{A} \models \psi[a_1,\ldots,a_n,\vec{b}].$$

Let $A^* = \{a_1,\ldots,a_n\}$ and for $i=1,\ldots,k$ define $R_i^* = R_i \cap (A^*)^{\lambda(i)}$ (R_i is assumed to be a $\lambda(i)$-ary predicate). Let $\mathfrak{A}^* = \langle A^*,R_1^*,\ldots,R_k^*\rangle$, the substructure of $\mathfrak{A}$ generated by $\{a_1,\ldots,a_n\}$. Now by lemma 1.30 and (*) it follows that

$$\text{for all } \vec{b} \in A^* \quad \mathfrak{A}^* \models \psi[a_1,\ldots,a_n,\vec{b}].$$

So by definition 1.21 $\mathfrak{A}^* \models \forall\vec{y}\psi[a_1,\ldots,a_n]$ whence by lemma 1.22 $\mathfrak{A}^* \models \sigma$. This completes the proof that the $\leq n$ element structure $\mathfrak{A}^*$ is a model for σ. □

Corollary 1.32. *If the sentence* σ *is of the form* $\forall x_1 \cdots \forall x_n \exists \vec{y}\psi$ *where* ψ *contains no quantifiers, function symbols or constants then* σ *is universally valid iff* σ *is valid in all domains with at most n elements.*

Proof: Apply theorem 1.31 to $\neg\sigma$. □

The final definition in connection with the semantics of a first order language is a generalization of definition 1.27(b).

Definition 1.33. If $\Gamma \cup \{\phi\} \subseteq Form(L(\delta))$ then Γ *logically implies* ϕ (alternatively ϕ is a *logical consequence* of Γ) $\Gamma \models \phi$, if whenever $\mathfrak{A}$ is a realization of type δ and a a sequence in A such that $\mathfrak{A} \models_a \psi$ for every $\psi \in \Gamma$ then $\mathfrak{A} \models_a \phi$.

A logical consequence of a set of formulae Γ is a necessary semantic consequence of Γ on account of its logical form alone (and the formalization of truth in a given structure). A logical consequence of the empty set, $\emptyset$, is a universally valid formula and in this case we write $\models \phi$ (instead of $\emptyset \models \phi$). We write $\mathfrak{A} \models_a \Gamma$ if $\mathfrak{A} \models_a \psi$ for all $\psi \in \Gamma \subseteq Form(L)$.

Exercise 1.18. (a) Show that lemma 1.30 can be generalized if we omit the restriction that σ contains no constants; i.e. show if $\sigma \in Sent(L)$ is of the form $\exists x_1 \ldots . \exists x_n \forall \vec{y} \psi$ where ψ contains no quantifiers, no function symbols, but m distinct constants, then σ is satisfiable only if σ is satisfiable in a domain with at most $n+m$ elements.

(b) Show by an example that lemma 1.30 is no longer true if we omit the restriction that σ contain no function symbols.

Exercise 1.19. Show that $\{\phi\} \models \psi$ if and only if $\models \phi \rightarrow \psi$.

Exercise 1.20. Are the following formulae universally valid?

(a) $\forall y(\forall x(F(x) \rightarrow (F(y) \rightarrow G(x))) \rightarrow (H(a) \rightarrow (\forall x F(x) \rightarrow G(y))))$

(b) $\exists x \exists y(((F(x) \leftrightarrow G(y)) \vee \forall z H(z)) \rightarrow (H(x) \& (F(x) \rightarrow G(y))))$

Exercise 1.21. Suppose that for $i=1,\ldots,m+1$ $\phi_i \in Sent(L)$ is of the form $\forall \vec{x} \psi_i$ where ψ_i is quantifier-free and contains no function symbols or constants. Describe a procedure to determine if ϕ_{m+1} is a logical consequence of $\{\phi_1,\ldots,\phi_m\}$.

Exercise 1.22. (a) Suppose $\phi \in Sent\ L(\mathfrak{U})$ is *universal* (that is, ϕ is of the form $\forall \vec{x} \psi$ where ψ is quantifier free) and that $\mathfrak{U}^* \subseteq \mathfrak{U}$. Show that if $\mathfrak{U} \models \phi$ then $\mathfrak{U}^* \models \phi$.

(b) Suppose $\phi \in Sent\ L(\mathfrak{U})$ is *existential* (of the form $\exists \vec{x} \psi$ where ψ is quantifier-free) and that $\mathfrak{U} \subseteq \mathfrak{U}^*$. Show that if $\mathfrak{U} \models \phi$ then $\mathfrak{U}^* \models \phi$.

2
A FORMAL SYSTEM FOR THE PREDICATE CALCULUS

In the previous chapter mathematical structures, the objects of mathematical investigation, were considered. From a rigorous unambiguous definition of a relational structure it is possible to interpret familiar mathematical objects as examples of this type of structure. We can express in a formal first order language certain propositions that may or may not be true in a given structure. To continue the analysis of the mathematical study of structures we consider the notion of a 'mathematical proof'. From a basic set of axioms (defining a class of structures) the mathematician makes logical deductions to produce theorems. In this chapter a formal deduction system is defined which embraces certain types of logical reasoning whereby some first order formula is inferred from a given set of formulae as hypotheses. The formalization is deceptively simple and even apparently restrictive in the kinds of logical deduction permitted in the system. The system is consistent in the strong sense that a formula ϕ derivable from a set Γ of hypotheses is necessarily a logical consequence of Γ (theorem 2.6). The converse result, that all logical consequences of Γ are also derivable from Γ, is proved in Chapter 3. That the formal system is complete in this sense is perhaps surprising; it establishes that all intuitively 'logical' arguments can be derived from very few basic principles.

2.1. AXIOMS AND RULES FOR THE PREDICATE CALCULUS

There are finitely many axiom schemata and rules of inference for the predicate calculus which fall naturally into two groups. The first group, of propositional axioms and rules, contains six different schemata and one rule. We take as axioms:

A1. $\phi \to (\psi \to \phi)$;

A2. $(\phi \to (\psi \to \chi)) \to ((\phi \to \psi) \to (\phi \to \chi))$;

A3. $(\neg\phi \to \psi) \to ((\neg\phi \to \neg\psi) \to \phi)$;

A4a. $\phi \,\&\, \psi \to \phi$; A4b. $\phi \,\&\, \psi \to \psi$;

A5. $\phi \to (\psi \to \phi \,\&\, \psi)$;

for all ϕ, ψ, $\chi \in Form(L)$. The propositional rule of inference is:

R1. Modus ponens: ψ is an immediate consequence of ϕ, $\phi \to \psi$ (for any ϕ, $\psi \in Form(L)$).

The second group of schemata and second rule explicitly involve the universal quantifier. For all ϕ, $\psi \in Form(L)$

A6. $\forall x\phi(x) \to \phi(t)$ where t is a term free for x in ϕ;

A7. $\forall x(\phi \to \psi) \to (\phi \to \forall x\psi)$ where ϕ contains no free occurrences of x

are axioms. (In A6 ϕ may or may not contain free occurrences of x and may contain other free variables.) The quantifier rule is

R2. Generalization: $\forall x\phi$ is an immediate consequence of ϕ (where $\phi \in Form(L)$ and x is any variable).

Deductions are made from the axioms using the rules of inference. More formally, we have

<u>Definition 2.1.</u> Suppose $\Sigma \cup \{\phi\} \subseteq Form(L)$. A *derivation of* ϕ *from the* (possibly empty) *set* Σ is a finite sequence $\phi_1,\ldots,\phi_n$ of formulae such that $\phi = \phi_n$ and for each $i \leq n$

either (i) ϕ_i is an axiom

or (ii) $\phi_i \in \Sigma$

or (iii) for some $j,k < i$ ϕ_i is an immediate consequence of ϕ_j,ϕ_k according to R1

or (iv) for some $j < i$ ϕ_i is an immediate consequence of ϕ_j according to R2 and the quantified variable x does not occur free in Σ.

When there is a derivation of ϕ from Σ and $\Sigma^* \subseteq$ *Form*(L) is any set such that $\Sigma \subseteq \Sigma^*$ then we say ϕ is derivable from Σ^* and write $\Sigma^* \vdash \phi$. ϕ is a *theorem of the predicate calculus* when there is a derivation of ϕ from the empty set, and in this case we write $\vdash \phi$ (instead of $\emptyset \vdash \phi$).

Axioms are the starting point for derivations (2.1(i)) and, as such, we require amongst other things that they be universally valid. The restrictions on the term t in A6 and on the variable x in A7 are designed to preclude precisely those instances of the schemata that are not always valid. They are perhaps most easily understood with specific examples in mind. The term y is *not* free for x in the formula $\exists y\, P(x,y)$ since the free occurrence of x lies within the scope of the quantifier $\forall y$ (cf. definition 1.17). Now we consider the formula:

$$\forall x\, \exists y P(x,y) \to \exists y P(y,y).$$

In the structure $\mathfrak{N} = \langle N, < \rangle$ with the set of natural numbers as domain and < as the binary relation corresponding to P the sentence $\forall x \exists y P(x,y)$ is valid since it states: 'For every natural number n there is a larger natural number m'; however $\exists y P(y,y)$ is manifestly false since it asserts that there is a natural number larger than itself.

Suppose now we consider the following unrestricted instance of A7:

$$\forall x(P(x) \to Q(x)) \to (P(x) \to \forall x Q(x)).$$

and the interpretation $\mathfrak{N} = \langle N, P, Q \rangle$ where N is the set of natural numbers, P is the unary relation '- is divisible by 4' and Q '- is even'. Then certainly $\mathfrak{N} \models \forall x(P(x) \to Q(x))$. However the formula $P(x) \to \forall x Q(x)$ has a free variable x and $\mathfrak{N} \models (P(x) \to \forall x Q(x))[n]$ will hold only when $\mathfrak{N} \models P[n]$ does not hold since we do not have $\mathfrak{N} \models \forall x Q(x)$. Hence if n is divisible by 4 then $\mathfrak{N} \models P[n]$ and so not $\mathfrak{N} \models (P(x) \to \forall x Q(x))[n]$; we can conclude that $\forall x(P(x) \to Q(x)) \to (P(x) \to \forall x Q(x))$ is not universally valid.

The restriction on the use of R2 in derivations (2.1(iv)) is also motivated by semantic considerations. The aim here is that a formula ϕ which is derivable from Σ should be a logical consequence of Σ. If we suppose that the set Σ contains the single formula $x = 0$ and consider the 'derivation' of $\forall x x = 0$ from Σ : $x = 0$, $\forall x x = 0$, not satisfying the restriction of 2.1(iv), then in the interpretation $\langle N, =, 0\rangle$ $x = 0$ is satisfied with x interpreted as 0 but $\forall x x = 0$ is false.

By inspecting definition 2.1 it can be seen that if $\Sigma \vdash \phi$ then there is some minimal, finite $\Sigma_0 \subseteq \Sigma$ such that $\phi_1, \ldots, \phi_n$, say, is a derivation of ϕ from Σ_0; namely, Σ_0 consists of those formulae ϕ_i in the derivation which do not occur as axioms ((i)) or as consequences of the rules ((iii),(iv)). The restriction of the use of R2 need only apply to that minimal Σ_0. Hence, for instance, if x does not occur free in ϕ then

$$\{\phi(y),\ \psi(x)\} \vdash \forall x(\psi(x) \rightarrow \phi(y)) \text{ is valid;}$$

the following is a derivation of $\forall x(\psi(x) \rightarrow \phi(y))$ from $\{\phi(y)\}$:

ϕ_1 : $\phi(y) \rightarrow (\psi(x) \rightarrow \phi(y))$ (instance of A1),

ϕ_2 : $\phi(y)$

ϕ_3 : $\psi(x) \rightarrow \phi(y)$ (from ϕ_1, ϕ_2 by R1),

ϕ_4 : $\forall x(\psi(x) \rightarrow \phi(y))$ (from ϕ_3 by R2).

So the free occurrence of x in $\psi(x)$ in $\{\phi(y), \psi(x)\}$ is not relevant in the particular derivation we have above.

Since the set *Form*(L) is infinite there are infinitely many instances of each axiom schema. However the infinite set of axioms is recursive;† that is, there is a decision

†A definition of 'recursive' may be found in any text book of recursive function theory, e.g. Rogers(1967).

procedure to determine whether or not a given $\phi \in Form(L)$ is an instance of A1-7. Similarly, there is a finite procedure to determine of a given finite sequence of formulae $\phi_1,\ldots,\phi_n$ whether it constitutes a derivation of ϕ from a recursive set $\Sigma \subseteq Form(L)$. For each $i=1,\ldots,n$ we simply determine whether or not ϕ_i is an axiom, ϕ_i is a member of Σ, ϕ_i follows from ϕ_j,ϕ_k by R1 for some j,k in the finite set $\{1,\ldots,i-1\}$, or ϕ_i follows from ϕ_j by R2 for some $j < i$ and the quantified variable does not occur free in those formulae of Σ occurring in $\{\phi_1,\ldots,\phi_{i-1}\}$.

If ϕ, ψ, χ are well formed formulae in a propositional language, built up from statement letters (propositional variables) using the propositional connectives &, $\neg$ then the formulae corresponding to the forms A1 - A5 are tautologies. It is useful to classify the formulae of a first order language L which have the form of a propositional tautology.

Definition 2.2. $\phi \in Form(L)$ is an *instance of a tautology* if there is a tautology ϕ' with distinct statement letters $P_1,\ldots,P_n$ and $\phi_1,\ldots,\phi_n \in Form(L)$ such that ϕ results from ϕ' when for each $i=1,\ldots,n$ each occurrence of P_i is replaced by ϕ_i.

Exercise 2.1. Which of the following formulae are instances of tautologies?

(a) $\forall x\phi(x) \rightarrow (\exists y\psi(y) \rightarrow \forall x\phi(x))$,

(b) $(\phi(x,y) \;\&\; \psi(x,y)) \rightarrow (\phi(x,y) \vee \chi(x))$,

(c) $\phi(x) \rightarrow \exists x\phi(x)$.

Lemma 2.3. *Each instance of a tautology is universally valid.*

Proof: Suppose ϕ is an instance of a tautology.

If $\mathfrak{A}$ is a realization for L and a a sequence in A

then $\mathfrak{A} \models_a \psi_1 \,\&\, \psi_2$ iff $\mathfrak{A} \models_a \psi_1$ and $\mathfrak{A} \models_a \psi_2$
and $\mathfrak{A} \models_a \neg\psi$ iff not $\mathfrak{A} \models_a \psi$ (definition 1.21).

A comparison with the truth tables for & and ¬

P	*Q*	*P*&*Q*
T	*T*	*T*
T	*F*	*F*
F	*T*	*F*
F	*F*	*F*

P	¬*P*
T	*F*
F	*T*

shows that the definition of $\models$ for the propositional connectives mimics the definition of truth value. More precisely, suppose $P_1,\ldots,P_n$ are the statement letters occurring in the tautology ϕ' from which ϕ results by the substitution of ϕ_i for P_i; if v is the assignment of truth values to propositional formulae generated by the definition: $v(P_i) = T$ if and only if $\mathfrak{A} \models_a \phi_i$, then for all subformulae ψ of ϕ built up from $\{\phi_1,\ldots,\phi_n\}$ using &, ¬ $\mathfrak{A} \models_a \psi$ if and only if $v(\psi') = T$. Since ϕ' is a tautology (i.e. $v(\phi') = T$ for all assignments of truth values v) then $\mathfrak{A} \models_a \phi$ for all realizations $\mathfrak{A}$ and all sequences a. □

Alternative sets of schemata yielding as axioms instances of tautologies could be used in place of A1 - 5. One aim in the formalization is to generate as theorems all universally valid formulae. So, as a corollary to lemma 2.3 we wish all instances of tautologies to be provable. A way to ensure that this is the case is to take as axioms in place of those generated by the schemata A1 - 5, *all* instances of tautologies. The set of all tautologies is, like the set of instances of A1 - 5, a recursive set since the truth table method can be used to determine whether a given formula is an instance of a tautology. However, as a corollary to the completeness theorem for the propositional calculus with, as axioms, all instances of A1 - 5 and modus ponens as the only rule we have the following lemma:

Lemma 2.4. *Each instance of a tautology ϕ is a theorem of the predicate calculus. Moreover there is a derivation $\phi_1,\ldots,\phi_n$ of ϕ (from the empty set) such that for $i \le n$*

either (i) ϕ_i is an axiom of the form A1 - 5

or (ii) ϕ_i is an immediate consequence of ϕ_j,ϕ_k for some $j,k < i$ by R1.

Proof: Suppose ϕ is an instance of a tautology resulting from the propositional tautology ϕ' when $\psi_1,\ldots,\psi_m \in Form(L)$ are substituted for the distinct statement letters $P_1,\ldots,P_m$ in ϕ'. The completeness theorem for the propositional calculus[†] implies that there is a derivation $\phi'_1,\ldots,\phi'_n$ of ϕ' in the propositional calculus; i.e. $\phi' = \phi'_n$ and for each $i \le k$ either ϕ'_i is a propositional instance of A1 - 5 or for some $j,k < i$, ϕ'_i follows from ϕ'_j,ϕ'_k by modus ponens. Let $\phi_i \in Form(L)$ be the result of replacing for $s=1,\ldots,m$ each occurrence of P_s in ϕ'_i by ψ_s and each statement letter $Q \neq P_1,\ldots,P_m$ by an arbitrary $\phi_Q \in Form(L)$ (for $i=1,\ldots,n$). Then $\phi_1,\ldots,\phi_n$ will be a derivation of ϕ in the required form. □

Exercise 2.2. Suppose that $\phi \in Form(L)$ is derivable and that there is a derivation $\phi_1,\ldots,\phi_n$ of ϕ of the form given in Lemma 2.4. Show that ϕ is an instance of a tautology.

[†] In a system with $\neg,\rightarrow$ as primitive connectives, A1-3 alone, together with modus ponens, are sufficient to generate all tautologies (see e.g. Mendelson (1964), page 36). However with $\neg,\&$ as primitive and $P \rightarrow Q$ *defined* by $\neg(P \& \neg Q)$ the additional axioms A4,5 are necessary. (The independence of A4,5 from A1-3 can be demonstrated using a 3-valued interpretation). The completeness of our system can be proved using the Kalmar method (as in Mendelson(1964), page 36 with & replacing $\rightarrow$ in the induction step of lemma 1.12). An alternative proof will be given at the end of Chapter 3.

It is precisely the complete character of the schemata A1 - 5 together with R1 for generating all instances of tautologies that is important. Any (recursive) set of schemata which together with the rule R1 is sufficient to generate as theorems all instances of tautologies could be used in place of A1 - 5. Throughout the remainder of this book we will assume the completeness of the set A1 - 5 (in the sense described above). However it will be clear when a given proof rests on this assumption. The completeness proof for the predicate calculus given in the following chapter can be adapted to provide a proof of the completeness of the schemata A1 - 5 for the propositional calculus (without, of course, *assuming* that all tautologies are derivable). This method is outlined in theorem 3.28.

It might at first sight seem possible that all universally valid formulae are instances of tautologies. That this is not the case is an immediate corollary of the following lemma.

<u>Lemma 2.5.</u> *Each instance of A6 and A7 is universally valid.*

Proof: (i) Let $\phi \in Form(L)$ be of the form $\forall x\psi(x) \rightarrow \psi(t)$ where t is a term free for x in ψ. Suppose $\mathfrak{A}$ is a realization for L and a a sequence of elements of the domain of $\mathfrak{A}$. By lemma 1.22(b) $\mathfrak{A} \models_{\mathsf{a}} \phi$ iff when $\mathfrak{A} \models_{\mathsf{a}} \forall x\psi(x)$ then $\mathfrak{A} \models_{\mathsf{a}} \psi(t)$. So suppose $\mathfrak{A} \models_{\mathsf{a}} \forall x\psi(x)$. Then for all elements $b \in A$ $\mathfrak{A} \models_{\mathsf{a}(b/_i)} \psi(x)$ where x is v_i. In particular $\mathfrak{A} \models_{\mathsf{a}(b/_i)} \psi(x)$ when $b = t^{\mathfrak{A}}[\mathsf{a}]$. Then by lemma 1.26 $\mathfrak{A} \models_{\mathsf{a}} \psi(t)$. This implies the universal validity of ϕ.

(ii) Suppose $\phi \in Form(L)$ is of the form $\forall x(\psi \rightarrow \chi) \rightarrow (\psi \rightarrow \forall x\chi)$ where there are no free occurrences of x in ψ. Suppose $\mathfrak{A} \models_{\mathsf{a}} \forall x(\psi \rightarrow \chi)$ where $\mathfrak{A}$ is some realization for L. Then for all $b \in A$ $\mathfrak{A} \models_{\mathsf{a}(b/_i)} \psi \rightarrow \chi$ (assuming x is v_i). Hence if $\mathfrak{A} \models_{\mathsf{a}(b/_i)} \psi$ then $\mathfrak{A} \models_{\mathsf{a}(b/_i)} \chi$. By theorem 1.23 since x is not free in ψ either $\mathfrak{A} \models_{\mathsf{a}(b/_i)} \psi$ for all $b \in A$ or it is not the case that $\mathfrak{A} \models_{\mathsf{a}} \psi$. Hence

$\mathfrak{A} \models_{\mathrm{a}} \psi$ implies that $\mathfrak{A} \models_{\mathrm{a}(b/_i)} \chi$ for all $b \in A$. Then $\mathfrak{A} \models_{\mathrm{a}} \psi \to \forall x\chi$. So for any realization $\mathfrak{A}$ of L and any sequence a, $\mathfrak{A} \models_{\mathrm{a}} \phi$. □

It is clear that an instance of A6 or A7 is not an instance of a tautology and yet by lemma 2.5 is universally valid. Thus in order to construct a formal system in which all universally valid formulae may be derivable it is necessary to have as axioms formulae which are not instances of tautologies or rules other than modus ponens (or both). What is perhaps surprising is that the addition of only two schemata A6 and A7 together with the rule of generalization is sufficient to generate all universally valid formulae.

2.2. CONSISTENCY.

The following theorem demonstrates that the system is consistent with the semantic concept of logical consequence. It is sometimes known as the soundness theorem.

Theorem 2.6. *Suppose* $\{\phi\} \cup \Sigma \subseteq Form(L)$. *If* $\Sigma \vdash \phi$ *then* $\Sigma \models \phi$.

Proof: We suppose $\Sigma_0 \vdash \phi$ for some minimal $\Sigma_0 \subseteq \Sigma$ and use induction on the length of the derivation $\phi_1,\ldots,\phi_n$ of ϕ from Σ_0 to show that for each $i \leq n$ $\Sigma_0 \models \phi_i$. From lemmas 2.3 and 2.5 it follows that when ϕ_i is an instance of A1 - 7 then $\models \phi_i$. So in this case clearly $\Sigma_0 \models \phi_i$. Trivially if $\phi_i \in \Sigma_0$ then $\Sigma_0 \models \phi_i$. Now suppose that for all $j < i$ $\Sigma_0 \models \phi_j$ and that ϕ_i follows from ϕ_{j_1} and $\phi_{j_2} = \phi_{j_1} \to \phi_i$ by R1 where $j_1, j_2 < i$. Suppose further that $\mathfrak{A}$ is a realization for L such that $\mathfrak{A} \models_{\mathrm{a}} \Sigma_0$. By the induction hypothesis $\mathfrak{A} \models_{\mathrm{a}} \phi_{j_1}$ and $\mathfrak{A} \models_{\mathrm{a}} \phi_{j_2}$. Then by lemma 1.22 $\mathfrak{A} \models_{\mathrm{a}} \phi_i$. So by definition 1.33 $\Sigma_0 \models \phi_i$. Finally suppose ϕ_i is $\forall v_k \phi_j$ and that it follows from ϕ_j by R2 where $j < i$. By definition 2.1 (iv) and the minimality of Σ_0, v_k does not occur free in Σ_0. Then for any realization $\mathfrak{A}$ for L and any sequence a in A

$\mathfrak{A} \models_a \Sigma_0$ implies $\mathfrak{A} \models_{a(b/_k)} \Sigma_0$ for all $b \in A$. Hence by the induction hypothesis (since $j < i$) if $\mathfrak{A} \models_a \Sigma_0$ then $\mathfrak{A} \models_{a(b/_k)} \phi$ for all $b \in A$. Hence $\mathfrak{A} \models_a \Sigma_0$ implies $\mathfrak{A} \models_a \forall v_k \phi_j$ and so $\Sigma_0 \models \phi_i$. To complete the proof note that since $\Sigma_0 \subseteq \Sigma$, if $\mathfrak{A} \models_a \Sigma$ then $\mathfrak{A} \models_a \Sigma_0$. Hence $\Sigma \models \phi$. □

The method of proof used in theorem 2.6, induction on the length of a derivation, is, like the type of induction first used in theorem 1.23, characteristic of many proofs of theorems about the formal system. A derivation is necessarily of finite length and furthermore each formula in the derivation arises in one of only four ways, as an axiom, as a member of Σ, or as a consequence of R1 or R2. Thus for the induction step all that is necessary is to check that the hypothesis still holds for a formula occurring in the derivation which has arisen by one of these four methods. This method of proof can be used to yield theorems concerning derivable formulae or derivations in any formal system in which the number of axiom schemata and rules of inference is finite. (If there are infinitely many schemata given in a uniform manner then again the induction method is applicable.)

Definition 2.7. A (possibly empty) $\Sigma \subseteq Form(L)$ is *consistent* if for no $\phi \in Form(L)$ is it the case that $\Sigma \vdash \phi$ and $\Sigma \vdash \neg\phi$. Σ is *inconsistent* if Σ is not consistent.

Corollary 2.8. *The empty set is consistent.*

Proof: Suppose there is some $\phi \in Form(L)$ such that $\vdash \phi$ and $\vdash \neg\phi$. Then by theorem 2.6 $\models \phi$ and $\models \neg\phi$ which is clearly impossible because it contradicts definition 1.21. □

Definition 2.9. A *theory* T in L is a set of sentences of L which is deductively closed, i.e. for each $\sigma \in Sent(L)$ $T \vdash \sigma$ if and only if $\sigma \in T$.

Lemma 2.10. *$\Sigma \subseteq Form(L)$ is inconsistent if and only if for each $\phi \in Form(L)$ $\Sigma \vdash \phi$.*

Proof: One way round the implication is obvious because of definition 2.7. Conversely suppose that Σ is inconsistent. Then for some $\phi \in Form(L)$ $\Sigma \vdash \phi$ and $\Sigma \vdash \neg\phi$. For any $\psi \in Form(L)$ the formula $\phi \to (\neg\phi \to \psi)$ is an instance of a tautology and hence by lemma 2.4 $\Sigma \vdash \phi \to (\neg\phi \to \psi)$. Two uses of the rule R1 yield $\Sigma \vdash \psi$ as required. □

Corollary 2.11. *A theory T in L is consistent iff $T \neq Sent(L)$.*

The following theorem demonstrates that there is a derived rule of the formal system which acts as a kind of converse to modus ponens. Use of this derived rule often shortens considerably formal derivations in the system (although, of course, it does not add to the power of the system).

Theorem 2.12. The Deduction Theorem.

If $\Sigma \cup \{\phi,\psi\} \subseteq Form(L)$ and $\Sigma \cup \{\phi\} \vdash \psi$ then $\Sigma \vdash \phi \to \psi$.

Proof: If $\Sigma \cup \{\phi\} \vdash \psi$ then either $\Sigma \vdash \psi$ or whenever $\Sigma_0 \vdash \psi$ for some $\Sigma_0 \subseteq \Sigma \cup \{\phi\}$ then $\phi \in \Sigma_0$. In the first case, since $\Sigma \vdash \psi \to (\phi \to \psi)$ (by A1) it follows by modus ponens that $\Sigma \vdash \phi \to \psi$. In the second case by induction on the length of the derivation $\psi_1,\ldots,\psi_n$ of ψ from a minimal $\Sigma_0 \cup \{\phi\} \subseteq \Sigma \cup \{\phi\}$ we show that for each $i \leq n$ $\Sigma_0 \vdash \phi \to \psi_i$. If ψ_i is an axiom or in Σ_0 then since $\psi_i \to (\phi \to \psi_i)$ is an instance of A1 and $\Sigma_0 \vdash \psi_i$ then $\Sigma_0 \vdash \phi \to \psi_i$. If ψ_i is itself ϕ then $\phi \to \psi_i$ is an instance of a tautology and hence provable by lemma 2.4. Now suppose ψ_i follows from ψ_j and $\psi_k = \psi_j \to \psi_i$ by R1 when $j,k < i$. By the induction hypothesis $\Sigma_0 \vdash \phi \to \psi_j$ and $\Sigma_0 \vdash \phi \to (\psi_j \to \psi_i)$. The formula $(\phi \to (\psi_j \to \psi_i)) \to ((\phi \to \psi_j) \to (\phi \to \psi_i))$ is an instance of A2 and so two applications of R1 yield $\Sigma_0 \vdash \phi \to \psi_i$. Finally suppose ψ_i is $\forall x \psi_j$ where $j < i$ and x is not free in

$\Sigma_0 \cup \{\phi\}$ (from definition 2.1). By the induction hypothesis $\Sigma_0 \vdash \phi \to \psi_j$; then R2 ($x$ not free in Σ_0) yields $\Sigma_0 \vdash \forall x(\phi \to \psi_j)$. Then using A7 we obtain $\Sigma_0 \vdash \phi \to \forall x \psi_j$ which is $\Sigma_0 \vdash \phi \to \psi_i$. This completes the proof. □

Because of the restrictions on the use of the rule R2 in a derivation of ψ from $\Sigma \cup \{\phi\}$ it is possible to 'carry over' ϕ from left to right across the symbol $\vdash$ (as proved in theorem 2.12). Some systems allow unrestricted use of the generalization rule in derivations. Then the statement of both the deduction theorem and theorem 2.6 must be modified. These modifications are cumbersome in comparison with the restriction on R2 in derivations. Another reason for preferring the restricted rule of generalization has been given on page 43.

<u>Exercise 2.3.</u> (a) Show that $\{\phi(x_1,\ldots,x_n)\}$ is consistent iff $\{\exists x_1 \ldots . \exists x_n \phi(x_1,\ldots,x_n)\}$ is consistent.

(b) Show that $\Sigma \subseteq Form(L)$ is consistent iff every finite subset of Σ is consistent.

<u>Exercise 2.4.</u> Show that if ϕ is built up from the formulae $\phi_1,\ldots,\phi_k$ using only the propositional connectives and that for $i=1,\ldots,k \vdash \phi_i \leftrightarrow \psi_i$ then $\vdash \phi \leftrightarrow \psi$ where ψ results when ψ_i is substituted for ϕ_i in ϕ.

<u>Exercise 2.5.</u> <u>The independence of axioms A6,7.</u>
Let *, ′ be the functions from $Form(L)$ to itself generated by:

(i) $\phi^* = \phi' = \phi$ if ϕ is atomic;

(ii) $(\neg\phi)^* = \neg(\phi^*), (\neg\phi)' = \neg(\phi')$;

(iii) $(\phi_1 \mathbin{\&} \phi_2)^* = \phi_1^* \mathbin{\&} \phi_2^*, (\phi_1 \mathbin{\&} \phi_2)' = \phi_1' \mathbin{\&} \phi_2'$;

(iv) $(\forall x\phi)^* = \forall\vec{y}\phi^*(\vec{y})$ where $\vec{y}$ consists of those variables occurring free in ϕ^*,

$(\forall x\phi)' = \phi'(c)$ where c is a (fixed) new constant substituted for the free occurrences of x in ϕ'.

Prove that

(a) if ϕ is derivable using A1-6, R1, R2 then $\vdash\phi^*$;

(b) if ϕ is derivable using A1-5, A7, R1, R2 then $\vdash\phi'$.

By considering instances ϕ,ψ such that $\vdash\phi$, $\vdash\psi$ and ϕ^*,ψ' are not universally valid show that

(c) A7 is independent of A1-6, R1, R2

(d) A6 is independent of A1-5, A7, R1, R2.

2.3. LANGUAGE AND METALANGUAGE.

In this chapter two essentially different types of theorem occur in the text. For one type, theorems in the formal system, there is a very precise definition (2.1) and a symbolic notation '$\vdash$' so that '$\vdash \phi$' means 'ϕ is derivable from a precisely defined set of axioms and two rules of inference, modus ponens and generalization'. On the other hand some theorems are prefaced by 'Theorem' (followed by a number), stated in English and proved using the kind of reasoning ordinarily accepted by the mathematician. Such theorems are sometimes called metatheorems in order to distinguish them from the formal theorems. For example, consider the deduction theorem, 2.12. The statement of the theorem involves quantification over sets of formulae; formulae and derivations in the system become objects of discourse. The proof of theorem 2.12 is accomplished using the principle of induction. This kind of reasoning is applicable because a derivation in the system is of finite length and the dependence of some stage in the proof on previous stages is clearly defined. The principle is justified on the basis of our intuitive understanding of the natural numbers. The working mathematician in general accepts the method of induction as a valid form of reasoning. Although the principle of induction has no immediate analogue in the formal system other kinds of argument have direct counterparts in the system and yet are justified for distinctly different reasons. For example, in corollary 2.8 reductio ad absurdum is used.

From the assumption that the corollary does not hold a contradiction results. So the corollary must hold. The formal analogue may be stated as follows: if $\neg\phi \vdash \psi$ and $\neg\phi \vdash \neg\psi$ then $\vdash \phi$. This is a derived rule of the formal system. By the deduction theorem if $\neg\phi \vdash \psi$ and $\neg\phi \vdash \neg\psi$ then $\vdash \neg\phi \rightarrow \psi$ and $\vdash \neg\phi \rightarrow \neg\psi$. Then by A3 and modus ponens $\vdash \phi$. The justification of this derived rule *in* the system depends on the axioms and rules of the system and not upon their intended interpretation.

It is important to distinguish between the two distinct languages, the formal first order language *L* and the metalanguage in which the formal system can be analysed. We have chosen English as the metalanguage although in principle we could equally well have adopted a formalized metalanguage. Similarly although we have not analysed the valid forms of reasoning in the metatheory it would have been possible to formalize the permitted forms of inference (such as induction and reductio ad absurdum). However such a formalization on the metalevel provides no justification for the metatheory. Indeed a meta-meta language is necessary to 'translate' such a metalanguage into one understood by the reader. The use of English as a metalanguage accentuates the distinction between the two levels, the formal system itself and the informal system in which the formal system is analysed, and does not require translation. The most important differences are emphasized in the following table:

Informal level	Formal level
(for analysis *of* the formal system.)	(for analysis within the formal system.)
English	First order language
Theorem	$\vdash$
Proof	Derivation
Rules of proof accepted by mathematicians	Axioms A1-7 Rules R1, R2

2.4. FURTHER METATHEORY

Some proper subsets of *Form*(*L*) play significant roles in both proof-theoretic and model-theoretic analysis. We shall be concerned with sets *S* with the following property:

(*) There is a map $*$: *Form*(*L*) $\to$ *S* such that $\vdash \phi$ iff $\vdash \phi^*$

or with the stronger property

(**) There is a map $**$: *Form*(*L*) $\to$ *S* such that $\vdash \phi \leftrightarrow \phi^{**}$.

Proper subsets with the property (**) are particularly useful; for any $\phi \in$ *Form*(*L*) there is a $\phi^{**} \in S$ which is proof-theoretically and (by theorem 2.6) semantically equivalent to ϕ. Hence a property of *Form*(*L*) relating to the proof theory of the formal system or to the model theory has only to be established for *S* in order for it to hold for all *Form*(*L*). Sets with the property (*) are less important in the metatheory but nevertheless are sometimes useful in considerations of the theorems of the system.

<u>Lemma 2.13.</u> *Sent*(*L*) *has the property* (*). *For any* $\phi(x_1,\ldots,x_n) \in$ *Form*(*L*)

$$\vdash \phi(x_1,\ldots,x_n) \quad \text{iff} \quad \vdash \forall x_1 \ldots\ldots \forall x_n \phi(x_1,\ldots,x_n)$$

Proof: If $\vdash \phi$ then by n applications of R2 it follows that $\vdash \forall x_1 \ldots\ldots \forall x_n \phi$.
Conversely if $\vdash \forall x_1 \ldots\ldots \forall x_n \phi$ then, since x_i is free for x_i in any $\psi \in$ *Form*(*L*), by A6 and R1 $\vdash \phi$. □

<u>Exercise 2.4.</u> Show that there exists $\phi \in$ *Form*(*L*) such that it is not the case that $\vdash \phi \leftrightarrow \forall x_1 \ldots, \forall x_n \phi$.

<u>Definition 2.14.</u> $\phi \in$ *Form*(*L*) is said to be in *prenex normal form* if it is of the form $Q_1 x_1 \ldots . Q_n x_n \psi$ where for $i=1,\ldots,n$ Q_i is a quantifier $\forall$ or $\exists$ and ψ contains no quantifiers.

PNF(*L*) is the subset of *Form*(*L*) consisting of all formulae in prenex normal form.

PNF(L) has the strong property (**) and moreover given any $\phi \in Form(L)$ there is an effective method to determine a $\phi^{**} \in PNF(L)$ provably equivalent to ϕ.

Theorem 2.15. *For each $\phi \in Form(L)$ there is a $\phi^{**} \in PNF(L)$ such that $\vdash \phi \leftrightarrow \phi^{**}$.*

Proof: The following formulae are all derivable:

(i) $\neg\forall x\psi \leftrightarrow \exists x \neg\psi$

(ii) $\neg\exists x\psi \leftrightarrow \forall x \neg\psi$

(iii) $\forall x\phi_1 \;\&\; \phi_2 \leftrightarrow \forall x(\phi_1 \;\&\; \phi_2)$ where x is not free in ϕ_2;

(iv) $\phi_1 \;\&\; \forall x\phi_2 \leftrightarrow \forall x(\phi_1 \;\&\; \phi_2)$ where x is not free in ϕ_1;

(v) $\exists x\phi_1 \;\&\; \phi_2 \leftrightarrow \exists x(\phi_1 \;\&\; \phi_2)$ where x is not free in ϕ_2;

(vi) $\phi_1 \;\&\; \exists x\phi_2 \leftrightarrow \exists x(\phi_1 \;\&\; \phi_2)$ where x is not free in ϕ_1.

By definition $\exists x \neg\psi$ is $\neg\forall x\neg\neg\psi$. $\psi \leftrightarrow \neg\neg\psi$ is an instance of a tautology. Hence $\forall x\psi \vdash \neg\neg\psi$ using A6, A1 and R1. Then by R2 $\forall x\psi \vdash \forall x\neg\neg\psi$. A similar argument shows that $\forall x\neg\neg\psi \vdash \forall x\psi$. Then by the deduction theorem and the definition of $\leftrightarrow$ and A5 we have $\vdash \neg\forall x\psi \leftrightarrow \exists x\neg\psi$. Similarly $\vdash \neg\exists x\psi \leftrightarrow \forall x\neg\psi$. $\forall x\phi_1 \;\&\; \phi_2 \vdash \phi_1$ using A4 and A6. Also $\forall x\phi_1 \;\&\; \phi_2 \vdash \phi_2$. So using A5 and R2 to generalize over x (by hypothesis not free in $\forall x\phi_1 \;\&\; \phi_2$) we obtain $\forall x\phi_1 \;\&\; \phi_2 \vdash \forall x(\phi_1 \;\&\; \phi_2)$. Conversely, $\forall x(\phi_1 \;\&\; \phi_2) \vdash \phi_1 \;\&\; \phi_2$ and hence $\forall x(\phi_1 \;\&\; \phi_2) \vdash \phi_1$ and $\forall x(\phi_1 \;\&\; \phi_2) \vdash \phi_2$. Using R2 to generalize over x and again A5 $\forall x(\phi_1 \;\&\; \phi_2) \vdash \forall x\phi_1 \;\&\; \phi_2$. The deduction theorem and the tautology $(A\rightarrow B) \rightarrow ((B\rightarrow A) \rightarrow (A\leftrightarrow B))$ imply that $\vdash \forall x(\phi_1 \;\&\; \phi_2) \leftrightarrow \forall x\phi_1 \;\&\; \phi_2$. Since $(A\&B) \leftrightarrow (B\&A)$ is a tautology we can readily show that $\vdash \phi_1 \;\&\; \forall x\phi_2 \leftrightarrow \forall x(\phi_1 \;\&\; \phi_2)$ if x is not free in ϕ_1. To prove (v) we show that, assuming x is not free in ϕ_2, both $\neg\exists x(\phi_1 \;\&\; \phi_2) \vdash \neg(\exists x\phi_1 \;\&\; \phi_2)$ and

$\neg(\exists x\phi_1 \;\&\; \phi_2) \vdash \neg\exists x(\phi_1 \;\&\; \phi_2)$. Then we use the deduction theorem and the tautology $(\neg A \leftrightarrow \neg B) \rightarrow (A \leftrightarrow B)$ to deduce $\vdash \exists x\phi_1 \;\&\; \phi_2 \leftrightarrow \exists x(\phi_1 \;\&\; \phi_2)$. By (ii) $\neg\exists x(\phi_1 \;\&\; \phi_2) \vdash \forall x\neg(\phi_1 \;\&\; \phi_2)$. Using the tautologies $\neg(A\&B) \leftrightarrow (\neg A \vee \neg B)$ and $(\neg A \vee \neg B) \leftrightarrow (B \rightarrow \neg A)$ we get $\neg\exists x(\phi_1 \;\&\; \phi_2) \vdash \forall x(\phi_2 \rightarrow \neg\phi_1)$. Then by A7, (ii) above and the second tautology $\neg\exists x(\phi_1 \;\&\; \phi_2) \vdash \neg\exists x\phi_1 \vee \neg\phi_2$. So $\neg\exists x(\phi_1 \;\&\; \phi_2) \vdash \neg(\exists x\phi_1 \;\&\; \phi_2)$. $\neg\exists x\phi_1 \;\&\; \phi_2 \vdash \neg\exists x(\phi_1 \;\&\; \phi_2)$ is proved in a similar way. (vi) is an almost immediate corollary of (v) just as (iv) follows from (iii).

Now suppose ϕ is written in its unabbreviated form with only the connectives and quantifier, $\neg$, &, and $\forall$. Using (i), (ii), (iii), or (iv) in 2.15 quantifiers can be 'pulled out' from right to left. When the universal quantifier is pulled across $\neg$ it is replaced by an existential quantifier. For this reason we include (ii), (v), and (vi) to show how to deal with the existential as well as the universal quantifier. We illustrate the procedure by an example thus showing how to obtain $\phi^{**} \in PNF(L)$ given $\phi \in Form(L)$. Suppose ϕ is

$$(\neg\forall x\mathrm{P}(x,y) \;\&\; \mathrm{Q}(x)) \;\&\; \neg(\forall z\mathrm{R}(z) \;\&\; \mathrm{S}(w))$$

where P, Q, R, and S are predicates of L. By (i)

$$\vdash \neg\forall x\mathrm{P}(x,y) \leftrightarrow \exists x\neg\mathrm{P}(x,y).$$

We cannot apply (v) directly to the formula $\exists x\neg\mathrm{P}(x,y) \;\&\; \mathrm{Q}(x)$ because x is free in Q. However it is straightforward to show that $\vdash \exists x\psi(x) \leftrightarrow \exists u\psi(u)$ where u is a variable not occurring in ψ (cf. the semantic parallel in lemma 1.25). So now by (v) we have $\vdash \exists x\neg\mathrm{P}(x,y) \;\&\; \mathrm{Q}(x) \leftrightarrow \exists u(\neg\mathrm{P}(u,y) \;\&\; \mathrm{Q}(x))$. By (iii) and (i) $\vdash \neg(\forall z\mathrm{R}(z) \;\&\; \mathrm{S}(w)) \leftrightarrow \exists z\neg(\mathrm{R}(z) \;\&\; \mathrm{S}(w))$. So by (v) $\vdash \phi \leftrightarrow \exists u\exists z(\neg\mathrm{P}(u,y) \;\&\; \mathrm{Q}(x) \;\&\neg(\mathrm{R}(z) \;\&\; \mathrm{S}(w)))$. □

Exercise 2.6. Show that

(a) $\vdash (\forall x\phi(x) \rightarrow \psi) \leftrightarrow \exists x(\phi(x) \rightarrow \psi)$ if x is not free in ψ,

(b) $\vdash (\exists x\phi(x) \to \psi) \leftrightarrow \forall x(\phi(x) \to \psi)$ if x is not free in ψ,
(c) $\vdash (\psi \to \forall x\phi(x)) \leftrightarrow \forall x(\psi \to \phi(x))$ if x is not free in ψ,
(d) $\vdash (\psi \to \exists x\phi(x)) \leftrightarrow \exists x(\psi \to \phi(x))$ if x is not free in ψ,
(e) $\vdash (\forall x\phi(x) \vee \psi) \leftrightarrow \forall x(\phi(x) \vee \psi)$ if x is not free in ψ,
(f) $\vdash (\exists x\phi(x) \vee \psi) \leftrightarrow \exists x(\phi(x) \vee \psi)$ if x is not free in ψ.

Exercise 2.7. Find prenex normal forms equivalent to the following formulae:
(a) $(\forall xP(x,y) \to Q(z)) \vee \exists z(R(z) \to \forall yS(x,y,z))$
(b) $\neg(\exists xP(x)\ \&\ \forall y(P(y) \vee Q(y,z)) \to \exists wR(y,z,w))$.

Exercise 2.8. Show that theorem 2.15 may be strengthened by replacing $PNF(L)$ with $PNF^*(L)$ where $PNF^*(L)$ contains precisely those formulae of the form $Q_1x_1 \ldots . Q_nx_n\psi$ where ψ is the disjunction of conjunctions of the form $\rho_1 \& \ldots . \& \rho_k$ where for each $i=1,\ldots,k$ ρ_i is either an atomic formula or the negation of an atomic formula.

Exercise 2.9. Show that $\vdash \phi(t) \to \exists x\phi(x)$ if t is free for x in $\phi(x)$.

Exercise 2.10. Show that the following are derived rules of the system:
(a) If $\Gamma, \phi(c) \vdash \psi$ where the constant c does not occur in Γ, ψ or $\phi(x)$
then $\Gamma, \exists x\phi(x) \vdash \psi$;
(b) If $\Gamma \vdash \phi(c)$ where the constant c does not occur in Γ, $\phi(x)$
then $\Gamma \vdash \forall x\phi(x)$.

Exercise 2.11. Show that $\vdash \exists x\forall y\phi(x,y) \to \forall y\exists x\phi(x,y)$ but that in general it is not the case that
$\vdash \forall y\exists x\phi(x,y) \to \exists x\forall y\phi(x,y)$.

Exercise 2.12. The Principle of Duality. Suppose $\phi \in Form(L)$ contains no function symbols and only the connectives $\vee$, &, $\neg$, $\forall$, $\exists$. Let ϕ^D be the formula which results from ϕ when $\vee$ is replaced by &, & by $\vee$, $\forall$ by $\exists$ and $\exists$ by $\forall$. Suppose $\mathfrak{A} = \langle A, R_1, \ldots, R_k, c_1, \ldots, c_l \rangle$ is a realization for the language of ϕ. Let $\mathfrak{A}^D = \langle A, R_1^D, \ldots, R_k^D, c_1, \ldots, c_l \rangle$ where for $i = 1, \ldots, k, R_i^D = A^{\lambda(i)} - R_i$.

Show that (a) $\mathfrak{A} \models \phi[\vec{a}]$ iff $\mathfrak{A}^D \models \neg \phi^D[\vec{a}]$

and (b) $\vdash \phi$ iff $\vdash \neg \phi^D$

2.5. FIRST ORDER THEORIES WITH EQUALITY.

The examples of relational structures given in Chapter 1 show that the appropriate first order language for a mathematical structure frequently contains a binary predicate corresponding to the equality relation. The most natural first order language used by working mathematicians for groups, rings, fields, arithmetic, and so on has the predicate = and functions instead of predicates representing functions. The intention is that = should be interpreted as identity in all realizations of the language. For the remainder of this section we consider a first order language with a distinguished binary predicate =. The language will be denoted by L_E and we write $t_i = t_j$ instead of $=(t_i, t_j)$ in order to follow standard mathematical notation more closely. When the language L_E is used the set of axioms of the corresponding formal system is extended to include the following:

A8. $\forall x(x = x)$;

A9. $x = y \rightarrow (\phi(x,x) \rightarrow \phi(x,y))$ for every atomic formula ϕ such that $\phi(x,y)$ is obtained from $\phi(x,x)$ by replacing x with y in some (not necessarily all) free occurrences of x in $\phi(x,x)$.

Note that we assume neither that ϕ has any free occurrences of x nor that x is the only free variable of ϕ. A8 is a single new axiom but the schema A9 generates an infinite set of axioms if the language L_E contains infinitely many non-logical symbols. It is immediately obvious that in any realization $\mathfrak{A}$ for L_E in which = is interpreted by equality the new axioms are valid and hence consistent. Unfortunately the converse, that the interpretation of = in any model of axioms

A1 - 9 is equality, is false. However, as a corollary to the following lemma, it is the case that in a model of A1 - 9 the interpretation of = is an equivalence relation on the domain of the structure.

If ϕ is derivable from Σ in a language L_E with equality (i.e. definition 2.1 with (i) meaning 'ϕ is an axiom of the kind A1-9') then we write $\Sigma \vdash_E \phi$. If $\mathfrak{A}$ is a realization for L_E we assume that the binary relation E representing = is such that A8 and each instance of A9 are valid in $\mathfrak{A}$.

Lemma 2.16. *(a)* $\vdash_E x = y \to y = x$

(b) $\vdash_E x = y \to (y = z \to x = z)$.

Proof: *(a)* Take $\phi(x,x)$ to be $x = x$ and $\phi(x,y)$ to be $y = x$ in A9. Then $\vdash_E x = y \to (x = x \to y = x)$ Now use the tautology $A \to ((B \to (A \to C)) \to (B \to C))$ and $\vdash_E x = x$ to obtain $\vdash_E x = y \to y = x$.

(b) Take $\phi(y,y)$ to be $y = z$ and $\phi(y,x)$ to be $x = z$ and use A9 with x and y interchanged to get $\vdash_E y = x \to (y = z \to x = z)$. By (i) and the tautology $(A \to B) \to ((B \to C) \to (A \to C))$ we obtain $\vdash_E x = y \to (y = z \to x = z)$ □

Corollary 2.17. *Suppose $\mathfrak{A}$ is a realization for L_E. Then E, the relation representing =, is an equivalence relation on A.*

Proof: E is reflexive since $\mathfrak{A} \models x = x$. The symmetry and transitivity of E are immediate consequences of lemma 2.16 (a) and (b) respectively. □

Although the number of instances of A9 may be finite (when the number of nonlogical symbols is finite and we identify formulae of the same form differing only in the free variables) there are infinitely many theorems with the same form as A9.

Theorem 2.18. *For all $\phi \in Form(L_E)$ $\vdash_E x = y \to (\phi(x,x) \to \phi(x,y))$ (in the notation of A9).*

Proof: (by induction on the length of ϕ). If ϕ is atomic then $x = y \rightarrow (\phi(x,x) \rightarrow \phi(x,y))$ is an axiom. Now suppose ϕ is of the form $\neg\psi$. Then by the induction hypothesis (with x,y interchanged) $\vdash_E y = x \rightarrow (\psi(x,y) \rightarrow \psi(x,x))$. The result $\vdash_E x = y \rightarrow (\neg\psi(x,x) \rightarrow \neg\psi(x,y))$ follows from lemma 2.16(a) and the propositional tautologies $(A \rightarrow B) \rightarrow (\neg B \rightarrow \neg A)$ and $(A \rightarrow B) \rightarrow ((B \rightarrow C) \rightarrow (A \rightarrow C))$. When ϕ is of the form $\phi_1 \,\&\, \phi_2$ the result comes immediately from the induction hypothesis and the tautology
$(A \rightarrow (B_1 \rightarrow C_1)) \rightarrow ((A \rightarrow (B_2 \rightarrow C_2)) \rightarrow (A \rightarrow (B_1 \& B_2 \rightarrow C_1 \& C_2)))$. If $\phi(x,x)$ is of the form $\forall z\, \psi(x,x,z)$ then by the induction hypothesis $\vdash_E x = y \rightarrow (\psi(x,x,z) \rightarrow \psi(x,y,z))$. Using R2 to generalize over z and then A7 (z not free in $x = y$) we obtain $\vdash_E x = y \rightarrow \forall z(\psi(x,x,z) \rightarrow \psi(x,y,z))$.
$\vdash \forall z(\mu \rightarrow \nu) \rightarrow (\forall z\, \mu \rightarrow \forall z\, \nu)$ is a simple consequence from A6, R1 and the deduction theorem. So combining this with the previous statement we have
$\vdash_E x = y \rightarrow (\forall z\, \psi(x,x,z) \rightarrow \forall z\, \psi(x,y,z))$. This completes the proof that for all $\phi \in Form(L_E)$ $\vdash_E x = y \rightarrow (\phi(x,x) \rightarrow \phi(x,y))$.□

A consequence of theorem 2.18 is that any model for a theory in L_E can be contracted to one in which = is interpreted by equality and the sets of sentences valid in the twc models are precisely the same.

<u>Definition 2.19.</u> Suppose $\mathfrak{A}$ is a realization for L_E. The *normal contraction* $\mathfrak{A}^N$ is defined as follows:

(i) $\mathfrak{A}^N$ is a structure of the same type as $\mathfrak{A}$.

(ii) A^N, the domain of $\mathfrak{A}^N$, is the set of equivalence classes of the domain of A induced by the equivalence relation E representing = in $\mathfrak{A}$. We write $[a]$ for the equivalence class containing $a \in A$.

(iii) If R_i is a relation in $\mathfrak{A}$ then R_i^N is defined by

$$\langle [a_1], \ldots, [a_{\lambda(i)}] \rangle \in R_i^N \quad \text{iff} \quad \langle a_1, \ldots, a_{\lambda(i)} \rangle \in R_i.$$

(iv) If f_j is a function in $\mathfrak{A}$ then f_j^N is defined by

$$f_j^N([a_1] \ldots [a_{\mu(j)}]) = [f_j(a_1 \ldots a_{\mu(j)})].$$

(v) If c_k is a distinguished element in $\mathfrak{A}$ then $c_k^N = [c_k]$.
$\mathfrak{A}^N$ is well defined (i.e. clauses (iii), (iv), and (v) are unambiguous) because

$$\mathfrak{A} \models_{\mathrm{a}} x = y \to (\mathrm{R}_i(t_1,\ldots,x,\ldots t_{\lambda(i)}) \to \mathrm{R}_i(t_1,\ldots,y,\ldots t_{\lambda(i)}))$$

and

$$\mathfrak{A} \models_{\mathrm{a}} x = y \to (\mathrm{f}_j(t_1,\ldots,x,\ldots t_{\mu(j)}) = (\mathrm{f}_j(t_1,\ldots,y,\ldots t_{\mu(j)})).$$

Definition 2.20. A realization $\mathfrak{A}$ for L_E is *normal* iff = is interpreted by equality.

Theorem 2.21. *If $\mathfrak{A}$ is a realization for L_E then the normal contraction $\mathfrak{A}^N$ is normal and for any $\phi(\vec{x}) \in Form(L_E)$ $\vec{a} \in A$*

$$\mathfrak{A} \models \phi[\vec{a}] \quad iff \quad \mathfrak{A}^N \models \phi[[\vec{a}]]$$

Proof: By definition 2.19 (iii) $\langle [a_1],[a_2]\rangle \in E^N$ iff $\langle a_1,a_2\rangle \in E$ where E is the interpretation of = in $\mathfrak{A}$. But $\langle a_1,a_2\rangle \in E$ iff $[a_1] = [a_2]$. So E^N is the equality relation in $\mathfrak{A}^N$, which is therefore normal.

To show that $\mathfrak{A} \models \phi[\vec{a}]$ iff $\mathfrak{A}^N \models \phi[[\vec{a}]]$ we show firstly by induction on the length of a term that for all $t \in Term(L_E)$ $[t^{\mathfrak{A}}[\mathrm{a}]] = t^{\mathfrak{A}^N}[[\mathrm{a}]]$. If t is v_i then $[t^{\mathfrak{A}}[\mathrm{a}]] = [a_i] = t^{\mathfrak{A}^N}[[\mathrm{a}]]$. If t is c_k then $[t^{\mathfrak{A}}[\mathrm{a}]] = [c_k] = t^{\mathfrak{A}^N}[[\mathrm{a}]]$. If t is $\mathrm{f}_j(t_1,\ldots,t_{\mu(j)})$ then by the induction hypothesis $[t_i^{\mathfrak{A}}[\mathrm{a}]] = t_i^{\mathfrak{A}^N}[[\mathrm{a}]]$ for all $i=1,\ldots,\mu(j)$. Then by definition 2.19 (iv) $[t^{\mathfrak{A}}[\mathrm{a}]] = t^{\mathfrak{A}^N}[[\mathrm{a}]]$.

Now we use induction on the length of ϕ to show that $\mathfrak{A} \models \phi[\vec{a}]$ iff $\mathfrak{A}^N \models \phi[[\vec{a}]]$
If ϕ is $\mathrm{R}_i(t_1,\ldots,t_{\lambda(i)})$ then

$$\mathfrak{A} \models \phi[\vec{a}] \quad \text{iff} \quad \langle t_1^{\mathfrak{A}}[\vec{a}],\ldots,t_{\lambda(i)}^{\mathfrak{A}}[\vec{a}]\rangle \in R_i$$

iff $\langle [t_1^{\mathfrak{A}}[\vec{a}]],\ldots,[t_{\lambda(i)}^{\mathfrak{A}}[\vec{a}]]\rangle \in R_i^N$ by definition

iff $\mathfrak{A}^N \models \phi[[\vec{a}]]$ by the above.

If ϕ is $\neg\psi$ or ϕ_1 & ϕ_2 then it follows immediately from the induction hypothesis that $\mathfrak{A} \models \phi[\vec{a}]$ iff $\mathfrak{A}^N \models \phi[[\vec{a}]]$. Lastly suppose ϕ is $\forall x\psi$. Then by definition 1.21 and the induction hypothesis

$\mathfrak{A} \models \phi[\vec{a}]$ iff for all $b \in A$ $\mathfrak{A} \models \psi[\vec{a},b]$

iff for all $[b] \in A^N$ $\mathfrak{A}^N \models \psi[[\vec{a}],[b]]$

iff $\mathfrak{A}^N \models \phi[[\vec{a}]]$. This completes the proof.□

<u>Corollary 2.22.</u> *If* $\Sigma \subseteq Form(L_E)$ *then* Σ *has a model iff* Σ *has a normal model.*

Proof: If Σ has a model $\mathfrak{M}$ then the normal contraction $\mathfrak{M}^N$ of $\mathfrak{M}$ is a normal model for Σ. □

From now on we will suppose that a realization for L_E is normal. Each realization $\mathfrak{A}$ for L_E is essentially the same as its normal contraction $\mathfrak{A}^N$ since precisely the same formulae are satisfiable in the two structures. Some authors do not consider non-normal realizations for languages with equality. One method to ensure that all realizations will be normal is to add the following clause to definition 1.21

$$\mathfrak{A} \models_{a} t_1 = t_2 \quad \text{iff} \quad t_1^{\mathfrak{A}}[a] = t_2^{\mathfrak{A}}[a].$$

It is important to remember though that = is, in the context of the language L_E, a purely formal symbol and without this additional clause (which ensures that axioms A8 and A9 are universally valid) need not be interpreted as equality.

Familiar mathematical theories can be axiomatized in a language with equality. For example, let $L_E(G)$ be the language with one binary function + and one constant 0. The theory of groups is then the set of theorems derivable from:

(i) $\forall x \forall y \forall z((x+y) + z = x + (y + z))$

(ii) $\forall x(x + 0 = x)$

(iii) $\forall x \exists y(x + y = 0)$.

If we add a new unary function - to the language $L_E(G)$ then (iii) can be replaced by the purely universal sentence (iii)' $\forall x(x + (-x) = 0)$. All (normal) models for (i), (ii), and (iii) will be groups as will those of (i), (ii), (iii)'. Models of (i), (ii), and (iii)' have the property that all substructures are subgroups by exercise 1.22(a). (The group axioms may also be formalized in a language L_E with + and no constants.)

Exercise 2.13. Show how to extend the axioms (i), (ii) and (iii)' for groups given above (in a suitably extended language) so that all models are

(a) rings and (b) fields

and that substructures are respectively rings and fields.

In the language L_E it is possible to formalize

'There is exactly one x such that $P(x)$'

where $P(x)$ is some proposition expressible in L_E. More precisely, suppose $\phi(x) \in Form(L_E)$. Let $\psi \in Form(L_E)$ be

$$\exists x(\phi(x) \,\&\, \forall y(\phi(y) \rightarrow x = y)).$$

Then if $\mathfrak{A}$ is a (normal) realization for L_E

$\mathfrak{A} \models \psi$ iff there is precisely one $a \in A$ such that $\mathfrak{A} \models \phi[a]$.

So we have

Definition 2.23. $\exists! x\phi(x) =_{Df} \exists x(\phi(x) \,\&\, \forall y(\phi(y) \rightarrow x = y))$.

Exercise 2.14. Show that for any realization $\mathfrak{M}$ for L_E

$$\mathfrak{M} \models \exists x \, \forall y(x=y \leftrightarrow \phi(y)) \leftrightarrow \exists! x \; \phi(x).$$

Exercise 2.15. Suppose

$\exists^2! x\phi(x) =_{Df} \exists x(\phi(x) \,\&\, \exists! y(\phi(y) \,\&\, \neg x = y))$.

Show that $\mathfrak{A} \models \exists^2! x\phi(x)$ iff there are precisely two distinct $a_1, a_2 \in A$ such that $\mathfrak{A} \models \phi[a_i]$ for $i=1,2$.

Exercise 2.16. By considering a suitable generalization of $\exists^2!x\phi(x)$ define a formula $\exists^n!x\phi(x)$ by induction on n such that $\mathfrak{A} \models \exists^n!x\phi(x)$ iff there are precisely n distinct

$a_1,\ldots,a_n \in A$ such that $\mathfrak{A} \models \phi[a_i]$ for $i=1,2,\ldots,n$.

Hence define a $\sigma_n \in Sent(L_E)$ such that

$\mathfrak{A} \models \sigma_n$ iff the domain of $\mathfrak{A}$ has exactly n elements.

In a language with equality associated with a (normal) structure $\mathfrak{A}$ it is possible to express properties of $\mathfrak{A}$ concerning the cardinality of some subsets of A. The preceding exercises show that 'There are precisely n elements of $\mathfrak{A}$ with the (first order) property ϕ' can be expressed in the language for $\mathfrak{A}$. The following lemma illustrates that the notions of a normal model and a language with equality are necessary for this kind of expressive power in a first order language.

Lemma 2.24. *Suppose $\mathfrak{A}$ is a realization of cardinality λ for the first order language L. For each cardinal $\kappa \geq \lambda$ there is a realization $\mathfrak{B}$ of cardinality κ such that $\mathfrak{A} \subseteq \mathfrak{B}$ and for any $\phi \in Form(L)$, $a_1,\ldots,a_n \in A$*

$$\mathfrak{A} \models \phi[a_1,\ldots,a_n] \quad \text{iff} \quad \mathfrak{B} \models \phi[a_1,\ldots,a_n]$$

Proof: First we extend A, the domain of $\mathfrak{A}$, to a set B with cardinality κ. Let a_0 be some fixed element of A. We define a map $* : B \to A$ as follows: $b^* = b$ if $b \in A$, $b^* = a_0$ if $b \notin A$. We shall now arrange that each element $b \in B$ has exactly those properties of its image under * in A. More precisely, suppose that for $i \in I$, R_i^* is the interpretation of R_i in $\mathfrak{B}$ and $b_1,\ldots,b_{\lambda(i)} \in B$. Then we define

$$\langle b_1,\ldots,b_{\lambda(i)}\rangle \in R_i^* \quad \text{iff} \quad \langle b_1^*,\ldots,b_{\lambda(i)}^*\rangle \in R_i.$$

Similarly the function f_j^* is defined on B as follows:

$$f_j^*(b_1,\ldots,b_{\mu(j)}) = f_j(b_1^*,\ldots,b_{\mu(j)}^*) \text{ for } b_1,\ldots,b_{\mu(j)} \in B, j \in J.$$

(R_i and f_j are respectively a relation and function in A.)
Since we want $\mathfrak{A} \subseteq \mathfrak{B}$ the constants c_k for $k \in K$ must necessarily have the same interpretation in the two structures.
It is now straightforward to show by induction on the length of $\phi \in Form(L)$ that for $b_1,\ldots,b_n \in B$

$$(*) \quad \mathfrak{B} \models \phi[b_1,\ldots,b_n] \text{ iff } \mathfrak{B} \models \phi[b_1^*,\ldots,b_n^*]$$
$$\text{iff } \mathfrak{A} \models \phi[b_1^*,\ldots,b_n^*].$$

For atomic formulae (*) is an immediate consequence of the definitions of R_i^* and f_j^*. For ϕ of the form $\neg\psi$ or ϕ_1 & ϕ_2 the result follows at once from the induction hypothesis.
When ϕ is $\forall x\psi(x,y_1,\ldots,y_n)$ then by the induction hypothesis for all $b,b_1,\ldots,b_n \in B$

$$\mathfrak{B} \models \psi[b,b_1,\ldots,b_n] \text{ iff } \mathfrak{B} \models \psi[b^*,b_1^*,\ldots,b_n^*]$$
$$\text{iff } \mathfrak{A} \models \psi[b^*,b_1^*,\ldots,b_n^*].$$

As b ranges over B, b^* ranges over A. Hence

$$\mathfrak{B} \models \phi[b_1,\ldots,b_n] \text{ iff } \mathfrak{B} \models \phi[b_1^*,\ldots,b_n^*]$$
$$\text{iff } \mathfrak{A} \models \phi[b_1^*,\ldots,b_n^*]. \qquad \square$$

<u>Corollary 2.25.</u> *If $\mathfrak{A}$ is a realization for L and $\mathfrak{A} \models \phi[a_0]$ for some $a_0 \in A$ then for any cardinal λ there is a realization $\mathfrak{B}$ such that*

$$\mathfrak{A} \subseteq \mathfrak{B} \quad \textit{and} \quad \{b : \mathfrak{B} \models \phi[b]\} \quad \textit{has cardinal} \geq \lambda.$$

Proof: Define the domain of $\mathfrak{B}$ to be $A \cup X$ where X is some set disjoint from A of cardinality λ. Define the map $* : B \to A$ as in the proof of lemma 2.24 and then define the relations, functions and constants of $\mathfrak{B}$ as before. We have then for all $b \in B$

$$\mathfrak{B} \models \phi[b] \text{ iff } \mathfrak{A} \models \phi[b^*]$$

So for all $b \in X$ $\mathfrak{B} \models \phi[b]$ (because $\mathfrak{A} \models \phi[a_0]$ and $b^* = a_0$). $\square$

A consequence of lemma 2.24 is that a structure for L (without equality) can trivially be expanded to one of arbitrarily large cardinality in which precisely the same formulae are valid. Since there is no upper bound on the size of a model of a satisfiable formula there is no syntactic way of limiting the size of a (class of) structures of a certain kind.

Exercise 2.17. Show that any model of ϕ where ϕ is

$$\forall x \neg P(x,x) \ \& \ \forall x \forall y \forall z(P(x,y) \ \& \ P(y,z) \rightarrow P(x,z)) \ \& \ \forall x \exists y P(x,y)$$

is necessarily infinite. Give examples of such models and show that the prenex normal form of ϕ cannot be of the form $\exists \vec{u} \forall \vec{v} \psi$ where ψ is quantifier free.

Although it is possible in a language with equality to define sentences whose models are of bounded finite cardinality, in the following chapter we shall show that even in such a language some concepts related to cardinality are not characterizable by a first order set of sentences.

3
THE COMPLETENESS THEOREM AND ITS COROLLARIES

In the previous chapter it was shown that the notion of formal derivability (in our system) is consistent with, and no stronger than, the notion of logical consequence. Our aim now is to show that the converse is also true and hence to deduce that the two concepts, (semantic) logical consequence and (syntactic) derivability, coincide. This was first proved by Godel (1930). The proof given here is essentially that due to Henkin (1949). The method is not only applicable to the particular system for the predicate calculus of Chapter 2 but can be generalized to establish the completeness of other formal systems. Other important metatheorems follow as corollaries to the completeness theorem and the particular form of the proof. One form of the completeness theorem relating the syntactic and semantic properties of a set of sentences is as follows:

<u>Theorem</u>. *A set* $\Sigma \subseteq Sent(L)$ *is consistent if and only if* Σ *has a model.*

(Other forms of the completeness theorem will be proved as corollaries.)

One implication, that a set of sentences with a model is necessarily consistent is an immediate consequence of theorem 2.6 since no realization can be a model both for some sentence σ and its negation $\neg\sigma$. Now we consider the converse implication. We suppose that $\Sigma \subseteq Sent(L)$ is a set known to have the (syntactic) property of being consistent. A model for Σ has somehow to be constructed. The tools we use are the actual symbols of the language of Σ (suitably extended if necessary to a richer language) which themselves determine a model for Σ.

3.1. DEFINITIONS AND OUTLINE OF THE PROOF.

<u>Definition 3.1.</u> $\Sigma \subseteq Sent(L)$ is *complete* if for each $\sigma \in Sent(L)$ either $\Sigma \vdash \sigma$ or $\Sigma \vdash \neg\sigma$.

It is important to realize that the notion of completeness of a set of sentences (or theory) is distinct from the notion of completeness of a formal system, which pertains to the equivalence between certain semantic and syntactic concepts. The completeness of Σ is, as stated here, entirely a syntactic property. However the completeness theorem for the predicate calculus implies that the property may equally well be defined in terms of semantic concepts.

<u>Definition 3.2.</u> A first order language L' is an *alphabetic extension* of the language L if L' is obtained from L by the addition of a set of new constant symbols alone. We write $L \subseteq_A L'$ when L' is an alphabetic extension of L.

<u>Definition 3.3.</u> Suppose $\Sigma \subseteq Sent(L)$ and $L \subseteq_A L'$. $\Sigma' \subseteq Sent(L')$ is a *Σ-full extension of* Σ in L' if $\Sigma \subseteq \Sigma'$ and for any formula $\phi(x) \in Form(L)$ with precisely one free variable x such that $\Sigma \vdash \exists x\phi(x)$ there is a constant $c \in L'$ such that $\phi(c) \in \Sigma'$.

<u>Definition 3.4.</u> $\Sigma \subseteq Sent(L)$ is *full* if Σ is a Σ-full extension of Σ in L.

If Σ' is a Σ-full extension of Σ in L' then an instantiating constant c corresponding to the derivable existential sentence $\exists x\phi(x)$ (i.e. $\Sigma' \vdash \exists x\phi(x) \rightarrow \phi(c)$) is sometimes called a Henkin constant.

We now outline the steps by which we show that a consistent set of sentences has a model.

A. Any consistent $\Sigma \subseteq Sent(L)$ can be embedded in a consistent Σ-full extension $\Sigma' \subseteq Sent(L')$ for some L'

such that $L \subsetneq_A L'$.

B. Any consistent $\Sigma \subseteq Sent(L)$ can be embedded in a complete consistent $\Sigma^* \subseteq Sent(L)$.

C. (Using A and B alternately) any consistent $\Sigma \subseteq Sent(L)$ can be embedded in a complete consistent full $\Sigma^+ \subseteq Sent(L^+)$ where $L \subsetneq_A L^+$.

D. A complete consistent full $\Sigma^+ \subseteq Sent(L^+)$ has a model.

E. (Using C and D) if $\Sigma \subseteq Sent(L)$ is consistent then has a model.

The method of proof involves the construction of a model (step D) which depends only on the sentences in Σ^+. An analysis of the construction of the model yields stronger theorems than that stated at the beginning of this chapter for we can ascertain bounds on the size of a canonical model associated with a consistent set of sentences. For languages L_E with equality the associated canonical model (D) is normal.

The first two lemmas in the proof, of which one is stated in A above, are valid for sets of sentences in arbitrary first order languages L. To continue the steps outlined above we consider separate cases according to the cardinality of the language L. The steps of the proof are the same in each case but the assumptions under which B can be carried out depend intrinsically on the cardinality of L.

<u>Lemma 3.5.</u> *Suppose* $\Sigma \subseteq Sent(L)$ *and* Σ *is consistent. Then there is an* L' *such that* $L \subsetneq_A L'$ *and a consistent* Σ*-full extension of* Σ, $\Sigma' \subseteq Sent(L')$.

Proof: There is an obvious way in which to extend Σ to a Σ-full set Σ' in a language with 'enough' constants. The problem is to show that the natural construction does yield a consistent Σ'. First we consider those existential sentences of L for which we require instantiating constants. Let

$\Sigma_1 = \{\phi \in Form(L)$: ϕ has exactly one free variable x_ϕ and $\Sigma \vdash \exists x_\phi \phi(x_\phi)\}$.

Now we simply add to L new constants not already occurring in L in one-one correspondence with the elements of Σ_1. That is, $L' = L \cup \{c_\phi : \phi \in \Sigma_1\}$ where $c_\phi \notin L$ and $c_\phi = c_\psi$ only if $\phi = \psi$. It is trivial that $\Sigma' = \Sigma \cup \{\phi(c_\phi) : \phi \in \Sigma_1\}$ is a Σ-full extension of Σ in L'. Now we must show that Σ' is consistent. We argue by contradiction. Suppose that Σ' is inconsistent. Then for some $\psi \in Form(L')$ we have both $\Sigma' \vdash \psi$ and $\Sigma' \vdash \neg\psi$. The derivations of ψ and $\neg\psi$ from Σ' are of finite length and hence there is a finite subset $\{\phi_1(c_{\phi_1}),\ldots,\phi_n(c_{\phi_n})\}$ of $\{\phi(c_\phi) : \phi \in \Sigma_1\}$ such that $\Sigma \cup \{\phi_i(c_{\phi_i})\}_{i=1}^n \vdash \psi$ and $\Sigma \cup \{\phi_i(c_{\phi_i})\}_{i=1}^n \vdash \neg\psi$. So the set $\Sigma \cup \{\phi_1(c_{\phi_1})\ \&\ \ldots\ \&\ \phi_n(c_{\phi_n})\}$ is inconsistent and, writing Φ for $\phi_1(c_{\phi_1})\ \&\ \ldots\ \&\ \phi_n(c_{\phi_n})$, we have $\Sigma \cup \{\Phi\} \vdash \neg\Phi$ by lemma 2.10. Then using the deduction theorem and the tautology $(\Phi \rightarrow \neg\Phi) \rightarrow \neg\Phi$, $\Sigma \vdash \neg\Phi$. The constants $c_{\phi_1},\ldots,c_{\phi_n}$ do not occur in Σ since by construction they are not in L, the language of Σ. So in the derivation of $\neg\Phi$ from Σ each occurrence of c_{ϕ_i} may be replaced by a distinct new variable $y_i \in L$ not already occurring in the derivation to yield a derivation of $\neg(\phi_1(y_1)\ \&\ \ldots\ \&\ \phi_n(y_n))$ from Σ. Using R2 to generalize over $y_1,\ldots,y_n$ we obtain $\Sigma \vdash \forall y_1.\ldots.\forall y_n\ \neg(\phi_1(y_1)\ \&\ \ldots\ \&\ \phi_n(y_n))$. Then, using the proof of theorem 2.15, $\Sigma \vdash \neg\exists y_1.\ldots.\exists y_n\,(\phi_1(y_1)\&\ \ldots\ \&\phi_n(y_n))$. Since the y_i's are distinct y_i does not occur in $\phi_j(y_j)$ if $i \neq j$ and it is easily shown that $\Sigma \vdash \neg(\exists y_1\phi_1(y_1)\ \&\ \ldots\ \&\ \exists y_n\phi_n(y_n))$. But since $\Sigma \vdash \exists y_i\phi_i(y_i)$ for $i=1,\ldots,n$ $(\phi_i \in \Sigma_1)$ this contradicts the consistency of Σ. Hence Σ' is consistent. □

<u>Lemma 3.6.</u> *Suppose* $\Sigma \cup \{\phi\} \subseteq Sent(L)$ *and it is not the case that* $\Sigma \vdash \neg\phi$. *Then* $\Sigma \cup \{\phi\}$ *is consistent.*

Proof: Notice that since $\neg\phi$ is not provable from Σ then by lemma 2.10 Σ is consistent. Suppose $\Sigma \cup \{\phi\}$ is inconsistent. Then $\Sigma \cup \{\phi\} \vdash \neg\phi$. Using the deduction theorem and the tautology $(\phi \rightarrow \neg\phi) \rightarrow \neg\phi$ we have $\Sigma \vdash \neg\phi$ which contradicts the hypothesis of the lemma. Hence $\Sigma \cup \{\phi\}$ is consistent. □

3.2. COMPLETENESS FOR COUNTABLE LANGUAGES

For this section we restrict our attention to sets of sentences in a countable language L. The proof of B (of the outline scheme in section 3.1) which we give for sets of sentences in a countable language does not generalize for the case where L is uncountable without a further set-theoretic axiom that is unnecessary in the countable case. The crucial property of a countable language L is that the set of sentences of L is itself countable and can be enumerated $\phi_1, \phi_2, \ldots$. The precise order in which the sentences occur in the list is unimportant; it is sufficient for our purposes that for each $\phi \in Sent(L)$ there is some finite n such that $\phi_n = \phi$. A listing of the sentences of L with this property can be obtained in the following way: list the symbols of the language (assumed countable) $s_1, s_2, \ldots$; at the nth stage consider the strings of symbols from $\{s_1, \ldots, s_n\}$ of length $\leq n$ (this is a finite set); eliminate those strings which are not sentences and those sentences already occurring in the list of sentences; then order the remaining (finite) set of sentences and add them to the list obtained up to the $(n-1)$th stage; then proceed to the $(n+1)$th stage. Each sentence in the language will automatically occur in this listing.

<u>Lemma 3.7.</u> *Suppose $\Sigma \subseteq Sent(L)$ where L is countable and Σ is consistent. Then there is a complete consistent $\Sigma^* \subseteq Sent(L)$ extending Σ.*

Proof: Suppose $\phi_1, \phi_2, \ldots$ is an enumeration of all the sentences of L. We define a countable sequence of sets of sentences $\Sigma_0, \Sigma_1, \ldots$ such that $\Sigma = \Sigma_0 \subseteq \Sigma_1 \subseteq \ldots\ldots \subseteq \Sigma_n \subseteq Sent(L)$

and Σ_n is consistent. Let $\Sigma_0 = \Sigma$. For $n \geq 0$ $\Sigma_{n+1} = \Sigma_n \cup \{\phi_{n+1}\}$ if it is not the case that $\Sigma_n \vdash \neg\phi_{n+1}$ and $\Sigma_{n+1} = \Sigma_n$ otherwise. Let $\Sigma^* = \bigcup_{n \in N} \Sigma_n$. The claim now is that each Σ_n is consistent and that Σ^* is complete and consistent. That Σ_n is consistent is proved by induction on n using lemma 3.6. If Σ^* is inconsistent then there is a finite $\Delta \subseteq \Sigma^*$ such that Δ is inconsistent (exercise 2.3(b)). But for any finite $\Delta \subseteq \Sigma^*$ there is an n such that $\Delta \subseteq \Sigma_n$ since if $\Delta = \{\sigma_1, \ldots, \sigma_m\}$ we may take $n = \max\{n_i : i = 1, \ldots, m\}$ where $\sigma_i \in \Sigma_{n_i}$. But this means Σ_n is inconsistent contradicting an earlier result in the proof. Consequently Σ^* is consistent.

It remains to show Σ^* is complete. Suppose $\sigma \in Sent(L)$ Then $\sigma = \phi_{n+1}$ for some $n \geq 0$. So either $\Sigma_n \vdash \neg\sigma$ or $\sigma \in \Sigma_{n+1}$. Hence Σ^* is complete. □

The particular complete consistent extension Σ^* of the consistent set Σ defined in lemma 3.7 is in no sense unique if Σ itself is not complete. For different orderings of the set of sentences different complete sets will result from the construction given in the proof. Indeed, for each $\sigma \in Sent(L)$ such that not $\Sigma \vdash \sigma$ and not $\Sigma \vdash \neg\sigma$, both $\Sigma \cup \{\sigma\}$ and $\Sigma \cup \{\neg\sigma\}$ are consistent and each can be extended to a complete consistent set including Σ.

In the following theorem we combine lemmas 3.5 and 3.7 in a suitable way to carry out step C in the outline scheme. When $\Sigma \subseteq Sent(L)$ is consistent and L is countable an analysis of the proof of lemma 3.5 shows that the lemma may be strengthened as follows: there is a countable alphabetic extension L' of L and a countable Σ-full extension $\Sigma' \subseteq Sent(L')$. L' is countable since Σ_1 is countable and the union of two countable sets is countable. Since Σ' is a set of sentences in a countable language L', Σ' is countable. We note that trivially a set of sentences in a countable language has a countable complete extension (since both sets are subsets of the countable set of sentences of the language).

Theorem 3.8. *Suppose $\Sigma \subseteq Sent(L)$ is consistent. Then there is an L^*, $L \subseteq_A L^*$ and a complete consistent full $\Sigma^* \subseteq Sent(L^*)$ extending Σ.*

Proof: We use lemmas 3.5 and 3.7 alternately countably often to construct (countable) languages $L_1, L_2, \ldots$ and (countable) sets of sentences $\Sigma_1, \Gamma_2, \Sigma_2, \Gamma_3, \ldots$ such that for each $i \geq 1$, $L_i \subseteq_A L_{i+1}$, $\Sigma_i \subseteq \Gamma_{i+1} \subseteq \Sigma_{i+1}$ and $\Gamma_{i+1} \cup \Sigma_{i+1} \subseteq Sent(L_{i+1})$. Let $L_1 = L$ and $\Sigma_1 = \Sigma$. When L_n and Σ_n have been defined then (by lemma 3.5) L_{n+1} and Γ_{n+1} are chosen so that $L_n \subseteq_A L_{n+1}$ and Γ_{n+1} is a consistent Σ_n-full extension of Σ_n in L_{n+1}. Σ_{n+1} is a complete consistent extension of Γ_{n+1} in L_{n+1} (lemma 3.7). Finally let $L^* = \bigcup_{n=1}^{\infty} L_n$ and $\Sigma^* = \bigcup_{n=1}^{\infty} \Sigma_n$. Σ^* is clearly an extension of Σ and $L \subsetneq_A L^*$. By construction each Σ_n is consistent and so, as in the proof of lemma 3.7 it follows that Σ^* is consistent. If $\sigma \in Sent(L^*)$ then for some $n \geq 1$ $\sigma \in Sent(L_n)$ and hence either $\Sigma_n \vdash \sigma$ or $\Sigma_n \vdash \neg\sigma$; so Σ^* is complete. To show Σ^* is full we suppose that $\Sigma^* \vdash \exists x\phi(x)$ where $\phi \in Form(L^*)$ and ϕ has the single free variable x. Then for some m, $\Sigma_m \vdash \exists x\phi(x)$ and so there is a constant $\mathrm{c} \in L_{m+1}$ such that $\phi(\mathrm{c}) \in \Gamma_{m+1} \subseteq \Sigma^*$. This completes the proof that Σ^* has the desired properties. □

We have now completed A, B, and C of the proof for countable languages L. Before we continue to D we need one further definition.

Definition 3.9. If $\Sigma \subseteq Sent(L)$ is consistent and L contains at least one constant the *canonical structure determined by* Σ is $\mathfrak{T}_\Sigma = \langle \tilde{T}, \{R_i\}_{i\in I}, \{f_j\}_{j\in J}, \{c_k\}_{k\in K}\rangle$

where $\tilde{T} = \{\tilde{t}: t$ is a closed (variable free) term of $L\}$

and $\langle \tilde{t}_1, \ldots, \tilde{t}_{\lambda(i)}\rangle \in R_i$ iff $\Sigma \vdash \mathrm{R}_i(t_1, \ldots, t_{\lambda(i)})$ for $i \in I$

$$f_j(\tilde{t}_1, \ldots, \tilde{t}_{\mu(j)}) = \widetilde{\mathrm{f}_j(t_1, \ldots, t_{\mu(j)})} \text{ for } j \in J$$

$$\text{and } c_k = \tilde{\mathrm{c}}_k \quad \text{for} \quad k \in K.$$

We here use a new notation $\tilde{t}$ for the term t to emphasize that the term assumes a new (semantic) role as an element in the domain of a relational structure and to distinguish this role from the original (syntactic) usage of t. So the set $\tilde{T}$ is simply a set of elements in one-one correspondence with the set of closed terms of L (non-empty since L is assumed to contain at least one constant).

The properties of completeness, fullness, and consistency are precisely those required to show that the canonical structure determined by Σ is indeed a model for Σ.

<u>Theorem 3.10.</u> Suppose $\Sigma \subseteq Sent(L)$ and that Σ is consistent, complete and full. For each $\sigma \in Sent(L)$

$$(*) \quad \mathfrak{T}_\Sigma \models \sigma \quad \text{iff} \quad \Sigma \vdash \sigma$$

Proof: We remark that if Σ if full then L has at least one constant (e.g. $\Sigma \vdash \exists x(R_i(x,x,\ldots,x) \vee \neg R_i(x,x,\ldots,x))$). We use induction on the length of $\sigma \in Sent(L)$ to prove (*). Note that we are using induction to establish a property of all *sentences* of L instead of the more familiar case of a property of all formulae of L. The induction steps must be modified accordingly: the atomic sentence σ is a formula of the form $R_i(t_1,\ldots,t_{\lambda(i)})$ where $t_1,\ldots,t_{\lambda(i)}$ are closed (variable-free) terms of L; $\neg\sigma$, σ_1 & σ_2 are sentences just in case σ, σ_1 and σ_2 respectively are sentences; $\forall x\phi(x)$ is a sentence if and only if for some (or, equivalently, all) closed term t of L $\phi(t)$ is a sentence and so at this stage of the induction the hypothesis can only be assumed for $\phi(t)$ where t is any closed term and not, as is usually the case, for the subformula $\phi(x)$. When σ is atomic then (*) is simply part of the definition of $\mathfrak{T}_\Sigma$.

Now suppose (*) holds for σ' when σ' has length less than that of σ. There are three cases to consider according to the outermost connective in σ.

1. σ is $\neg\sigma'$.

$\mathfrak{I}_\Sigma \models \sigma$ iff not $\mathfrak{I}_\Sigma \models \sigma'$ by definition 1.21

iff not $\Sigma \vdash \sigma'$ by the induction hypothesis

iff $\Sigma \vdash \neg\sigma'$ because Σ is consistent and complete.

2. σ is $\sigma_1 \,\&\, \sigma_2$.

$\mathfrak{I}_\Sigma \models \sigma$ iff $\mathfrak{I}_\Sigma \models \sigma_1$ and $\mathfrak{I}_\Sigma \models \sigma_2$ by definition 1.21

iff $\Sigma \vdash \sigma_1$ and $\Sigma \vdash \sigma_2$ by the induction hypothesis

iff $\Sigma \vdash \sigma_1 \,\&\, \sigma_2$ because $\sigma_1 \to (\sigma_2 \to (\sigma_1 \& \sigma_2))$ and $(\sigma_1 \& \sigma_2) \to \sigma_i$ for $i=1,2$ are instances of A4 and A5.

3. σ is $\forall x\, \phi(x)$.

$\mathfrak{I}_\Sigma \models \sigma$ iff for all $\tilde{t} \in \tilde{T}$ $\mathfrak{I}_\Sigma \models \phi[\tilde{t}]$ by definition 1.21

iff for all $t \in T^c$ (T^c is the set of closed terms) $\mathfrak{I}_\Sigma \models \phi(t)$ since the interpretation of the closed term t is $\tilde{t}$

iff for all $t \in T^c$ $\Sigma \vdash \phi(t)$ by the induction hypothesis since $\phi(t) \in Sent(L)$.

Suppose that it is not the case that $\Sigma \vdash \sigma$. Then by the completeness of Σ, $\Sigma \vdash \neg\sigma$. But $\Sigma \vdash \neg\sigma$ iff $\Sigma \vdash \exists x\, \neg\phi(x)$ (by theorem 2.15). Σ is full so for some $c \in L$ $\Sigma \vdash \neg\phi(c)$. Then by the equivalence above it is not the case that $\mathfrak{I}_\Sigma \models \sigma$. Conversely if $\Sigma \vdash \sigma$ then by A6 for all closed terms t $\Sigma \vdash \phi(t)$ and so $\mathfrak{I}_\Sigma \models \sigma$. This completes the induction and hence the theorem is established. □

A combination of theorems 3.8 and 3.10 will now complete the proof of the theorem for countable languages stated at the beginning of the chapter.

<u>Theorem 3.11.</u> *(Gödel-Henkin) Any consistent $\Sigma \subseteq Sent(L)$ for countable L has a model.*

Proof: By theorem 3.8 there is an alphabetic extension L^* of L and a complete consistent full $\Sigma^* \subseteq Sent(L^*)$ extending Σ. The reduct of $\mathfrak{I}_{\Sigma^*}$ to the language L is a model for Σ. □

Several important theorems are corollaries of theorem 3.11. The first asserts the completeness of the formal system in the form in which the notion was first formulated.

<u>Theorem 3.12.</u> *If $\Sigma \cup \{\sigma\} \subseteq Sent(L)$ then $\Sigma \models \sigma$ iff $\Sigma \vdash \sigma$.*

Proof: That $\Sigma \vdash \sigma$ implies $\Sigma \models \sigma$ was proved in theorem 2.6. Conversely suppose that it is not the case that $\Sigma \vdash \sigma$. Then it is also not the case that $\Sigma \vdash \neg\neg\sigma$. So by lemma 3.6 $\Sigma \cup \{\neg\sigma\}$ is consistent and by theorem 3.11 has a model $\mathfrak{I}$. This implies that not $\Sigma \models \sigma$. So $\Sigma \models \sigma$ iff $\Sigma \vdash \sigma$ demonstrating that formal derivability is as strong a concept as logical consequence. □

<u>Corollary 3.13.</u> *The theorems of the predicate calculus are precisely the universally valid formulae.*

Proof: Let Σ be empty and σ be the universal closure of some formula ϕ. Theorem 3.12 then implies that $\models \sigma$ iff $\vdash \sigma$. However ϕ is universally valid iff σ is universally valid (exercise 1.17(b)) and ϕ is derivable iff σ is derivable. Hence the result. □

<u>Theorem 3.14.</u> <u>Compactness theorem.</u> *If $\Sigma \subseteq Sent(L)$ then Σ has a model if and only if every finite subset of Σ has a model.*

Proof: If Σ has a model then trivially every finite subset of Σ has a model.

Conversely suppose Σ has no model. Then by theorem 3.11 Σ is inconsistent. So for some finite subsets $\Sigma_1, \Sigma_2 \subseteq \Sigma$ and some $\phi \in Form(L)$ $\Sigma_1 \vdash \phi$ and $\Sigma_2 \vdash \neg\phi$

Consequently the finite subset of Σ, $\Sigma_1 \cup \Sigma_2$ is inconsistent and hence has no model. □

The compactness theorem is formulated entirely in terms of semantic concepts. The theorem is meaningful without the apparatus of a formal system for first order logic. Indeed the theorem can be proved without reference to formal systems[†] but here it is an almost trivial consequence of the completeness theorem, which relates the syntactic and semantic ideas. The proof is totally non-constructive in the sense that there is no method for constructing a model for Σ given models for each finite subset of Σ. The compactness theorem is probably the most important and fundamental theorem in model theory and some applications will be given when we have established the completeness theorem for a larger class of first order languages L.

3.3. COUNTABLE LANGUAGES L_E WITH EQUALITY.

The construction of $\mathfrak{T}_\Sigma$ for complete consistent full sets of sentences Σ does not in general yield a normal model for Σ. However using the construction of theorem 2.21 it is immediately clear that a consistent set $\Sigma \subseteq Sent(L_E)$ has a normal model. In the case where L_E is a language with equality we will collapse the canonical structure associated with $\Sigma \subseteq Sent(L_E)$ so that it is automatically a normal model for Σ when Σ is complete, consistent and full.

<u>Definition 3.15.</u> If $\Sigma \subseteq Sent(L_E)$ and L_E contains at least one constant then the *canonical structure* determined by Σ is

$$\mathfrak{T}_\Sigma = \langle \overline{T}, \{R_i\}_{i\in I} \quad \{f_j\}_{j\in J} \quad \{c_k\}_{k\in K}\rangle$$

where $\overline{T}$ is the set of equivalence classes $[t]$ of closed terms t of L_E determined by the equivalence relation

---

[†]See e.g. Bell and Slomson (1969), page 102.

$$t_1 \sim t_2 \text{ iff } \Sigma \vdash t_1 = t_2$$

and (as before)

$$\langle [t_1],\ldots,[t_{\lambda(i)}]\rangle \in R_i \text{ iff } \Sigma \vdash \mathrm{R}_i(t_1,\ldots,t_{\lambda(i)}) \text{ for } i \in I$$

$$f_j([t_1],\ldots,[t_{\mu(j)}]) = [\mathrm{f}_j(t_1,\ldots,t_{\mu(j)})] \text{ for } j \in J$$

$$c_k = [\mathrm{c}_k] \text{ for } k \in K$$

The proof of theorem 3.10 can now be carried out as before to obtain a normal canonical model for complete consistent full sets of sentences in a countable L_E. Consequently

<u>Theorem 3.16A.</u> *If* $\Sigma \subseteq Sent(L_E)$ *and* L_E *is countable then* Σ *is consistent if and only if* Σ *has a (normal) model.*

As before, from this theorem may be deduced the analogues of corollary 3.13 and theorem 3.14 (the compactness theorem). One further important result may be proved at this stage. The proofs of theorems 3.8 and 3.10 are sufficiently constructive to consider the cardinality of a canonical model associated with a given countable consistent set of sentences. In theorem 3.8 a countable sequence of countable languages $\langle L_n \rangle_{n\in\omega}$ is constructed by adjoining at each stage a countable set of new constants. To ensure that the resulting language $L^* = \bigcup_{n\in\omega} L_n$ is countable also (without the axiom of choice) we may first define the language $L^+ = L \cup \{\mathrm{c}_i : i\in\omega\}$. Then let $f : \omega \to \omega \times \omega$ be some one-one onto function. At the nth stage in the construction instantiating constants for sentences of the form $\exists x\phi(x) \in Sent(L_n)$ such that $\Sigma_n \vdash \exists x\phi(x)$ are taken from the countably infinite set $\{\mathrm{c}_i : \exists m\ f(i) = \langle n,m\rangle\}$. Then the countable language L^+ is an alphabetic extension of each L_n and hence of L^*. Consequently the set of equivalence classes of closed terms of L^* is at most countable. Theorem 3.16A can be strengthened as follows:

Theorem 3.16B. *If* L_E *is countable and* $\Sigma \subseteq Sent(L_E)$ *then* Σ *is consistent if and only if* Σ *has a countable model.*

Of course theorem 3.16B is also valid for sets of sentences in a countable language without equality. But cardinality concepts are of considerably more significance for languages with equality (cf. pages 64,65).

The remainder of this section is concerned with some refinements of the completeness theorem and its proof for which familiarity with certain concepts of recursion theory and first order arithmetic is desirable. The reader may omit these paragraphs (which are not referred to elsewhere) and resume again at the beginning of Section 3.4.

A countable structure is isomorphic to a structure whose domain is a subset of N, the set of natural numbers. So, without loss of generality, the completeness theorem as proved in theorems 3.11 and 3.16 may be stated:

A countable consistent set of sentences is satisfiable in a subset of the natural numbers (with suitable relations, functions and individual constants).

Certain subsets of N (or subsets of N^n) admitting a definition of a particularly simple form are arithmetic; they can be classified further in the hierarchy of arithmetic sets of which those at the lowest level are the recursive sets. When a sentence is satisfied in a domain included in the natural numbers the interpretations for the predicate letters in the sentence are simply subsets of N^n and as such may be arithmetic. Kleene has shown that it is always possible to find a model (with domain $\subseteq N$) for a satisfiable sentence in which the relations are Δ^0_2 (relatively low in the arithmetic hierarchy); i.e. each n-ary relation R has the form

$$\{\langle x_1,\ldots,x_n\rangle \;:\; \forall y\, \exists z S_1(y,z,x_1,\ldots,x_n)\}$$

and the dual form

$$\{\langle x_1,\ldots,x_n\rangle \;:\; \exists y \forall z S_2(y,z,x_1,\ldots,x_n)\}$$

where S_1, S_2 are recursive.[†] Although some refinements of this result have been obtained it is not always possible to improve this result significantly; for example in general we cannot ensure that the relations are recursively enumerable, let alone recursive.

One of the corollaries of the completeness theorem is that each universally valid formula can be derived in the formal system of Chapter 2. A particular mathematical theory is obtained by considering the logical consequences of the axiom schemata A1 - 7 and additional non-logical axioms determined by the theory. Suppose now that instead of considering the addition of single instances of non-logical axioms we consider PC^+, the system resulting from the adjunction of a new axiom schema Aϕ with the form of an unprovable formula ϕ. Using the Gödel technique of arithmetization Hilbert-Bernays showed that first-order arithmetic (the formal system based on Peano's axioms) is ω-inconsistent when PC^+ is the underlying logic; i.e. there is some formula $\psi(x)$ such that $\psi(0)$, $\psi(1)$, $\psi(2)$,... and $\neg\forall x\psi(x)$ are all provable.[††] This is clearly an undesirable consequence of the underlying system for the predicate calculus; the schemata A1 - 7 (and their consequences using R1 and R2) are indeed sufficient.

3.4. COMPLETENESS FOR UNCOUNTABLE LANGUAGES

Suppose that L is a first order language of cardinality κ. If $\kappa \leq \aleph_0$ (i.e. is countable) then there is an enumeration of the countable set of sentences of L and, as a corollary, every consistent set of sentences of L can be extended to a complete consistent set. An analysis of the proof of lemma

---

[†] See Kleene (1962), page 394, Theorem 35.

[††] See Kleene (1962), page 395.

3.7 shows that the well-ordered enumeration of the sentences plays an integral part in the construction of the complete set extending the given consistent set. If κ is uncountable then in a standard formulation of set theory (e.g. Zermelo-Fraenkel or Gödel-Bernays) it is consistent to assume there is no such well ordering of the set of sentences. Hence in order to extend lemma 3.7 to uncountable languages it is necessary to add a new set theoretic axiom, the axiom of choice, to the metatheory. For the remainder of this section we will assume the axiom of choice and write (AC) beside theorems with proofs dependent upon this assumption. The specific form of the axiom of choice to be used is given in the following:

Well ordering theorem (AC). *Any set can be well ordered.*

A proof that the axiom of choice is actually equivalent to the proposition that any set can be well ordered can be found in any standard text book on set theory.†

A further corollary of the axiom of choice which will be used is stated in the following

Lemma 3.17 (AC). *Suppose κ is an infinite cardinal. Then* $\kappa = \kappa \cdot \aleph_0 = \kappa \cdot \kappa$.

A proof of this lemma can again be found in a standard set theory text.†

Corollary 3.18 (AC). *If L is a first order language with cardinality κ then the set of sentences of L has cardinality κ.*

Proof: There are $\leq \kappa^n$ sentences of L of length n since there are $\leq \kappa$ possible choices for each symbol. Hence there

†See e.g. Halmos (1960), pages 68, 97.

are $\leq \sum_{n\in\omega} \kappa^n = \sum_{n\in\omega} \kappa = \aleph_0 \cdot \kappa = \kappa$ sentences of L. That there are at least κ sentences follows from the observation that either there are κ distinct predicates or there are κ distinct terms of L. □

Lemma 3.7 can now be extended for an arbitrary first order language L.

Lemma 3.19 (AC). *Suppose $\Sigma \subseteq Sent(L)$ and Σ is consistent. Then there is a complete consistent $\Sigma' \subseteq Sent(L)$ such that $\Sigma \subseteq \Sigma'$.*

Proof: The set of sentences of L can be well-ordered $\{\phi_\alpha : \alpha < \beta\}$ where β is some ordinal with the same cardinality as L. By analogy with the proof of lemma 3.7 we define a sequence of sets of sentences of L as follows: $\Sigma_0 = \Sigma$; $\Sigma_{\alpha+1} = \Sigma_\alpha \cup \{\phi_\alpha\}$ if it is not the case that $\Sigma_\alpha \vdash \neg\phi_\alpha$ and $\Sigma_{\alpha+1} = \Sigma_\alpha$ otherwise; and for limit ordinals $\lambda \leq \beta$ $\Sigma_\lambda = \bigcup_{\alpha<\lambda} \Sigma_\alpha$. It follows from lemma 3.6 that if Σ_α is consistent then so also is $\Sigma_{\alpha+1}$. For limit λ, if Σ_λ is inconsistent then for some $\alpha_0 < \lambda$ Σ_{α_0} is inconsistent. So by transfinite induction Σ_α is consistent for each $\alpha \leq \beta$. Let $\Sigma' = \Sigma_\beta$. Σ_β is complete because for any $\sigma \in Sent(L)$ there is an $\alpha < \beta$ such that $\sigma = \phi_\alpha$. So either $\Sigma_\alpha \vdash \neg\phi_\alpha$ or $\phi_\alpha \in \Sigma_{\alpha+1} \subseteq \Sigma_\beta$. □

Exercise 3.1. Show that lemma 3.19 can be proved by applying Zorn's Lemma to the partially ordered set of consistent extensions of Σ in L.

The remainder of the proof of the completeness theorem is carried out as in the countable case. The countability of L is explicitly used only in the proof of step B which in the general case is lemma 3.19. The proof of theorem 3.8 is again accomplished by the application of lemmas 3.5 and 3.19 alternately countably often. Increasing sequences of sets $\langle\Sigma_n\rangle$, $\langle\Gamma_n\rangle$ are defined and then the union

$\Sigma^* = \bigcup_{n\in\omega} \Sigma_n$ is a complete consistent full extension of the consistent set Σ. The canonical structure is a model for Σ^* and hence its reduct to the language of Σ is a model for Σ. The generalized completeness theorem can be formulated as follows:

Theorem 3.20 (AC). *If $\Sigma \subseteq Sent(L)$ [$Sent(L_E)$], Σ is consistent and L has cardinality κ then Σ has a [normal] model of cardinality $\leq \kappa$.*

Proof: The cardinal bound on the canonical structure associated with a complete consistent full extension of Σ is obtained as follows:

(i) If L has cardinality $\leq \kappa$ then by corollary 3.18 $\Sigma_1 = \{\phi \in Form(L): \Sigma \vdash \exists x\phi(x)\}$ has cardinality $\leq \kappa$. Hence there is an alphabetic extension L' of L with cardinality $\leq \kappa + \kappa = \kappa$ and a Σ-full $\Sigma' \subseteq Sent(L')$ extending Σ.

(ii) If L has cardinality $\leq \kappa$ and $\Sigma \subseteq Sent(L)$ then there is a complete extension Σ^* of Σ with cardinality $\leq \kappa$.

(iii) The sets Σ_n, Γ_n for $n \in N$ (cf. proof of theorem 3.8) have cardinality $\leq \kappa$ and hence Σ^*, a complete consistent saturated extension of Σ, can be constructed in a language L^* with cardinality $\leq \kappa \cdot \aleph_0 = \kappa$. The set of equivalence classes of closed terms of L^* has cardinality $\leq \kappa$ establishing the required bound on the canonical structure. □

We have shown how to construct a model for a consistent set of *sentences*. What, if any, is the corresponding result for a consistent set of *formulae*, Φ? For simplicity, we will consider here only the case where Φ is a countable consistent set of *quantifier-free* formulae. Notice that the addition of Henkin constants when dealing with sets of sentences was necessary only for the quantifier case of the induction in theorem 3.10, so we would hope to be able to omit that step when considering quantifier-free formulae. The following theorem demonstrates how the result and the construction involved in the proof may be modified in this case.

Theorem 3.21. *If Φ is a countable consistent set of quantifier-free formulae of L then Φ is satisfiable in a countable structure.*

Proof: There are essentially two steps in the proof (corresponding to steps B and D of the outline of section 3.1). First we extend Φ to a consistent set Φ' complete with respect to quantifier-free formulae. A straightforward modification of the construction in lemma 3.7 accomplishes this:

namely, we enumerate the quantifier-free formulae of L, say $\phi_0, \phi_1, \ldots$ and define

$$\Phi_0 = \Phi; \; \Phi_{n+1} = \begin{cases} \Phi_n \cup \{\phi_n\} \text{ if not } \Phi_n \vdash \neg\phi_n; \\ \Phi_n \text{ otherwise.} \end{cases} \qquad \Phi' = \bigcup_{n\in\omega} \Phi_n$$

We leave it to the reader to verify that Φ' is consistent and that for any quantifier-free $\psi \in Form(L)$ either $\Phi' \vdash \psi$ or $\Phi' \vdash \neg\psi$.

Now construct a countable structure $\mathfrak{B}$ (depending on Φ') as follows:

(i) as domain, B, we take the set of *all* terms of L,

(ii) for $i \in I$ $\langle t_1, \ldots, t_{\lambda(i)} \rangle \in R_i$ iff $\Phi' \vdash \mathrm{R}_i(t_1, \ldots, t_{\lambda(i)})$,

(iii) for $j \in J$ $f_j(t_1, \ldots, t_{\mu(j)})$ is to be the term $\mathrm{f}_j(t_1, \ldots, t_{\mu(j)})$,

(iv) for $k \in K$ the interpretation of c_k is just c_k.

It is simple to verify that for any $t(x_1, \ldots, x_n) \in Term(L)$

$$t^{\mathfrak{B}}[x_1, \ldots, x_n] = t(x_1, \ldots, x_n)$$

(i.e. the denotation of t with respect to the assignment of the element x_i in B to the variable x_i is just t itself).

Now we show by induction on the length of a quantifier-free formula ϕ that

$$(\dagger) \quad \mathfrak{B} \models \phi[x_1,\ldots,x_n] \text{ iff } \Phi' \vdash \phi(x_1,\ldots,x_n).$$

(Notice that, as for the terms above, the $x_1,\ldots,x_n$ occurring on the left-hand side are to be thought of as elements in the domain of the structure $\mathfrak{B}$ whereas on the right-hand side they are terms in the language L.)

When $\phi(x_1,\ldots,x_n)$ is $R_i(t_1,\ldots,t_{\lambda(i)})$ (so that $x_1,\ldots,x_n$ are the variables occurring in the terms $t_1,\ldots,t_{\lambda(i)}$) then by definition 1.21

$$\begin{aligned}\mathfrak{B} \models \phi[x_1,\ldots,x_n] \quad &\text{iff} \quad \langle t_1,\ldots,t_{\lambda(i)}\rangle \in R_i \\ &\text{iff} \quad \Phi' \vdash R_i(t_1,\ldots,t_{\lambda(i)}) \text{ by definition.}\end{aligned}$$

The two induction steps corresponding to the propositional connectives $\neg$ and & are almost exactly as in theorem 3.10. Then since we are considering only quantifier-free formulae this completes the proof. □

<u>Exercise 3.2.</u> Suppose ϕ is a quantifier-free formula such that h terms occur in ϕ. Show that if ϕ is satisfiable then ϕ is satisfiable in a structure with $\leq h$ elements.

We have seen how the Henkin method for proving the completeness theorem in the countable case readily extends to the uncountable case. (This case was first proved by Malcev (1936).) This is not true of Gödel's original proof and is one reason for preferring the Henkin method of proof here. The completeness of certain systems other than first order predicate logic can also be established with a Henkin-style proof. Henkin (1950) himself used the method to prove a suitable formulation of the completeness theorem for for a system of order ω. Another application can be made in the case of infinitary logic as introduced by Barwise where the language $\mathscr{L}_A$ is based on a countable admissible set A.[†] Other methods for proving completeness include that of Rasiowa and Sikorski (1950), which uses Boolean algebras, and of Beth (1955) using semantic tableaux.

Although we have used the well ordering theorem, a set-theoretic principle equivalent to the axiom of choice, to establish the completeness theorem for arbitrary first order languages, a weaker principle, the Boolean prime ideal theorem, implied by, but not implying, the axiom of choice, is sufficient to show completeness in the following sense: a consistent first order theory has a model.[††]

3.5. APPLICATIONS OF THE COMPACTNESS THEOREM

In this section we begin with a general model-theoretic result before considering applications of the compactness theorem to particular mathematical theories.

Theorem 3.22.[†††] *Suppose* $\Sigma \subseteq Sent(L_E)$ *has arbitrarily large finite models. Then* Σ *has an infinite model.*

Proof: We define an alphabetic extension L_E^* of L_E by adjoining a countably infinite set of distinct new constants $\{b_i : i \in \omega\}$ to L_E. Let $\Sigma^* \subseteq Sent(L_E^*)$ be the union of Σ with the set of all sentences of the form $\neg b_i = b_j$ where $i < j < \omega$. Now suppose Σ_0 is a finite subset of Σ^*. Then $\Sigma_0 = \Sigma_1 \cup \{\neg b_{i_k} = b_{j_k}\}_{k=1}^{p}$ for some finite p and some finite $\Sigma_1 \subseteq \Sigma$. By hypothesis Σ has a model $\mathfrak{A}$ with more than $2p$ elements. We expand $\mathfrak{A}$ to a realization $\mathfrak{A}^*$ for L_E^* by adjoining distinct new constants $d_j \in A$ for the $\leq 2p$ new constant symbols b_j occurring in Σ_0 and arbitrary interpretations d_j for those b_j which do not occur in Σ_0. Then

---

[†] See e.g. Aczel (1973)

[††] See e.g. Mendelson (1964), page 100.

[†††] In this section the axiom of choice is assumed whenever necessary to yield the completeness theorem for uncountable languages.

$$\mathfrak{A}^* = \langle \mathfrak{A}, \{d_i : i \in \omega\}\rangle \models \Sigma_0.$$

Now we can apply the compactness theorem to the set $\Sigma^* \subseteq Sent(L_E^*)$. Since an arbitrary finite subset Σ_0 of Σ^* has a model so also does Σ^*, say $\mathfrak{B}^*$. $\mathfrak{B}^*$ has an infinite domain since there are infinitely many distinct elements that are the interpretations for the set $\{b_j\}_{j\in\omega}$. The reduct of $\mathfrak{B}^*$ to the language L_E is an infinite model for Σ. □

Corollary 3.23. *There is no set of sentences* Σ *such that* $\mathfrak{M} \models \Sigma$ *if and only if* $\mathfrak{M}$ *is finite.*

Corollary 3.23 can be contrasted with the result (exercise 2.16) that it is possible to characterize in a single sentence (in a language with equality) all (normal) structures with a given fixed finite cardinality, or indeed those structures with domains of less than some fixed finite cardinality. Finiteness itself cannot be characterized in the same way even when infinite sets of sentences are considered.

The compactness theorem can be used in more specific circumstances to obtain results in other branches of mathematics. This is illustrated with examples from algebra, arithmetic, and graph theory.

Theorem 3.24. *Suppose* L_E *is a first order language for the theory of fields with two binary functions* $+, \cdot$ *and two constants* $0, 1$. *The theory of fields of characteristic zero cannot be finitely axiomatized (i.e. there is no finite* $\Sigma \subseteq Sent(L_E)$ *such that* $\mathfrak{F} \models \Sigma$ *iff* $\mathfrak{F}$ *is a field of characteristic zero).*

Proof: The theory of fields can be axiomatized by the following set of sentences in L_E.

$$\text{I}\ \left\{\begin{array}{l} \forall x \forall y \forall z((x+y)+z = x+(y+z)) \\ \forall x \forall y(x+y = y+x) \\ \forall x(x+0 = x) \\ \forall x \forall \dot{y}(x+y = 0) \end{array}\right.$$

$$\text{II}\left\{\begin{array}{c}\forall x\,\forall y\,\forall z((x \cdot y) \cdot z = x \cdot (y \cdot z))\\ \forall x\,\forall y(x \cdot y = y \cdot x)\\ \forall x(x \cdot 1 = x)\\ \forall x(\neg x = 0 \to \exists y(x \cdot y = 1))\end{array}\right.$$

$$\text{III} \quad \forall x\,\forall y\,\forall z(x \cdot (y + z) = (x \cdot y) + (x \cdot z))$$

$$\text{IV} \quad \neg 0 = 1$$

Any model for group I will be an additive abelian group. A model for group II will, without the interpretation for 0, be a multiplicative group. III asserts that multiplication, $\cdot$, is distributive over addition, $+$. Let $F = \text{I} \cup \text{II} \cup \text{III} \cup \text{IV}$.

An axiomatization for fields of characteristic zero is obtained by adding to F an infinite set of sentences $\{S_n\}_{n\in\omega}$ where S_n is

$$\neg(\underbrace{1 + 1 + \dots \quad + 1}_{\text{n times}}) = 0$$

(A field is of characteristic zero if and only if it is not of non-zero finite characteristic.)

Let $F_0 = F \cup \{S_n\}_{n\in\omega}$.

Just because there is an axiomatization for fields of characteristic zero with an infinite set of axioms there is no a priori reason why a finite set $\Sigma \subseteq Sent(L_E)$ should not yield F_0 as theorems. The following lemma however eliminates this possibility.

<u>Lemma:</u> *Suppose* $\sigma \in Sent(L_E)$ *and* $F_0 \models \sigma$; *then there is an* n *such that* σ *is valid in all fields with characteristic* $\geq n$.

Proof: Suppose $F_0 \models \sigma$. Suppose there is no n such that σ is valid in all fields with characteristic $\geq n$. Then $\neg\sigma$ is satisfiable in a field with arbitrarily large finite characteristic. So any finite subset of $F_0 \cup \{\neg\sigma\}$ has a model.

So by compactness $F_0 \cup \{\neg\sigma\}$ has a model which contradicts the initial hypothesis that $F_0 \models \sigma$. Hence there is some n such that σ is valid in all fields with characteristic $\geq n$.

N.B. The proof of the lemma given above does not give us any *particular* n satisfying the lemma. The proof merely shows that the hypothesis that no such n exists leads to a contradiction. However if we use the completeness theorem (to replace $\models$ by $\vdash$) then from a *derivation* of σ from F_0 it is possible to compute an n: the derivation of σ will use only a finite number of instances of $\{S_m\}$ say, $S_{m_1},\ldots,S_{m_k}$; then (by completeness again) we conclude that for $n > max\{m_1,\ldots,m_k\}$ σ is valid in all fields of characteristic $\geq n$.

Now we return to the theorem. If $\psi_1,\ldots,\psi_k$ is a finite axiomatization for F_0 then consider $\sigma = \psi_1 \& \ldots \& \psi_k$. By the lemma σ is valid in all fields with characteristic $\geq n$ for some finite n. Since there are fields with arbitrarily large finite characteristic (e.g. J_p, the field of integers modulo p where p is prime) we have the required contradiction. □

<u>Exercise 3.3.</u> Assume that for any non-constant polynomial $p(x)$ with coefficients in some field $\mathfrak{F}$ there is an extension field $\mathfrak{F}^*$ of $\mathfrak{F}$ in which $p(x)$ has a root. Show, by using the compactness theorem, that for any field $\mathfrak{F}$ there is an extension field $\mathfrak{F}^*$ of $\mathfrak{F}$ in which every non-constant polynomial of $\mathfrak{F}[x]$ has a root.

The next example is an application of compactness to arithmetic.

<u>Theorem 3.25.</u> *Suppose L_E is a first order language with one unary function $'$ (successor), two binary functions $+$, $\cdot$, and one constant 0. Let $\mathfrak{N} = \langle N, ', +, \cdot, 0\rangle$ be the realization of L_E with domain N, the natural numbers, successor, addition and multiplication the interpretations for $'$, $+$, $\cdot$ respectively, and 0 the interpretation for 0. Let*

$Th(\mathfrak{N}) = \{\sigma : \sigma \in Sent(L_E)\ \&\ \mathfrak{N} \models \sigma\}$. *Then there is an* $\mathfrak{M}_0$ *such that* $\mathfrak{M}_0 \models Th(\mathfrak{N})$ *and* $\mathfrak{M}_0$ *is not isomorphic to* $\mathfrak{N}$.

Proof: Let L_E^+ be an alphabetic extension of L_E obtained by the addition of a single new constant c.

Let $\Sigma^+ = Th(\mathfrak{N}) \cup \{\sigma_n\}_{n\in\omega}$ where σ_n is $\neg \mathrm{c} = \mathrm{n}$ and

n is $0\underbrace{''\ldots'}_{n \text{ times}}$. Suppose Σ_0 is any finite subset of Σ^+.

Then there is some n_0 such that for all $m \geq n_0$ $\neg \mathrm{c} = \mathrm{m} \notin \Sigma_0$. Then $\langle \mathfrak{N}, n_0\rangle \models \Sigma_0$. Hence by the compactness theorem Σ^+ has a model, $\mathfrak{M}$ say. Suppose $\mathfrak{M}_0$, the reduct of $\mathfrak{M}$ to the language L_E, is isomorphic to $\mathfrak{N}$. Then there is an isomorphism $f : \mathfrak{N} \to \mathfrak{M}_0$. So for some integer n $f(n) = c$ where c is the interpretation of c in $\mathfrak{M}$. But

$$\mathfrak{M}_0 \models \neg x = \mathrm{n}[c] \text{ and } \quad \mathfrak{N} \models x = \mathrm{n}[n]$$

which gives the required contradiction. □

<u>Exercise 3.4.</u> Suppose $\mathfrak{M}_0$ is the (countable) model for $Th(\mathfrak{N}) \cup \{\sigma_n\}_{n\in\omega}$ occurring in the proof of theorem 3.24.

Define $\phi(x,y)$ to be $\exists z(\neg z = 0\ \&\ x + z = y)$

Then $\forall x \forall y(\phi(x,y) \vee \phi(y,x) \vee x = y) \in Th(\mathfrak{N})$.

Hence there is a linear ordering $<_{\mathfrak{M}_0}$ of $\mathfrak{M}_0$ defined by $a <_{\mathfrak{M}_0} b$ iff $\mathfrak{M}_0 \models \phi[a,b]$.

What is the order type of this ordering of the domain of $\mathfrak{M}_0$?

The final illustration of the compactness theorem is taken from graph theory.

<u>Definition 3.26.</u> (i) The structure $\langle M,R\rangle$ where R is a binary relation on M is a *graph* if R is irreflexive and symmetric on M. If $a,b \in M$ and $\langle a,b\rangle \in R$ then a is *connected* to b.

(ii) The graph $\langle M,R\rangle$ is *k-colourable* if there is a partition of M into $\leq k$ subsets such that no two connected elements are in the same subset.

Theorem 3.27. *Suppose* $\langle M,R\rangle$ *is a graph.* $\langle M,R\rangle$ *is k-colourable if and only if every finite substructure of* $\langle M,R\rangle$ *is k-colourable.*

Proof: Clearly if $\langle M,R\rangle$ is k-colourable then any finite substructure of $\langle M,R\rangle$ is k-colourable.

To prove the converse, that if every finite substructure of $\langle M,R\rangle$ is k-colourable then so also is $\langle M,R\rangle$, we introduce a first order language for $\langle M,R\rangle$ and add to it k unary predicates $\mathsf{P}_1,\ldots,\mathsf{P}_k$ and constants c_a for each $a \in M$. Let Δ be the following set of sentences of L:

$\mathsf{R}(\mathsf{c}_a,\mathsf{c}_b)$ for each pair $\langle a,b\rangle$ such that $\langle a,b\rangle \in R$

$\neg\mathsf{R}(\mathsf{c}_a,\mathsf{c}_b)$ for each pair $\langle a,b\rangle$ such that $\langle a,b\rangle \notin R$

$\forall x(\mathsf{P}_1(x) \vee \ldots\ldots \vee \mathsf{P}_k(x))$

$\forall x((\mathsf{P}_i(x) \rightarrow \neg\mathsf{P}_j(x))$ for each pair (i,j) with $1 \leq i\neq j \leq k$

$\forall x\,\forall y(\mathsf{P}_i(x)\ \&\ \mathsf{P}_i(y) \rightarrow \neg\mathsf{R}(x,y))$ for $i=1,\ldots,k$.

Δ has the same cardinality as M if M is infinite.

Now suppose every finite subgraph of $\langle M,R\rangle$ is k-colourable. Let $\Delta_0 \subseteq \Delta$ be a finite subset of Δ and suppose $M_0 = \{a_1,\ldots,a_p\}$ is the finite subset of M consisting of those elements a such that c_a occurs in Δ_0. By assumption $\langle M_0, R \cap (M_0 \times M_0)\rangle$ is k-colourable. So there is a partition of M_0 into k disjoint subsets, $S_1,\ldots,S_k$ say, such that for $i=1,\ldots,k$ if $a,b \in S_i$ then $\langle a,b\rangle \notin R$. Take some $a_0 \in M_0$. If a_0 is the interpretation c_a for $a \notin M_0$ then

$\langle M_0, R \cap (M_0{\times}M_0), S_1,\ldots,S_k, \{a\}_{a\in M_0} \cup \{c_a\}_{a\notin M_0}\rangle \models \Delta_0$. By compactness we obtain a model $\mathfrak{B} = \langle B, R^*, P_1,\ldots,P_k, \{c_a\}_{a\in M}\rangle$ for Δ in which P_i is the interpretation for P_i and c_a the interpretation for c_a. The substructure of $\mathfrak{B}$, say $\mathfrak{B}_0$, with domain $\{c_a\}_{a\in M}$ is also a model for Δ because each sentence in Δ is universal and hence valid in all substructures of models of Δ (exercise 1.22(a)). The structure $\langle B_0, R^*\rangle$ is isomorphic to $\langle M,R\rangle$ since $\mathfrak{B}_0 \models \mathsf{R}(c_a,c_b)$ if and only if

$\langle a,b\rangle \in R$. We define k disjoint subsets of M, $Q_1,\ldots,Q_k$ as follows: $a \in Q_i$ iff $c_a \in P_i \cap B_0$. Since $\mathfrak{B}_0 \models \forall x(\mathrm{P}_1(x) \vee \ldots \vee \mathrm{P}_k(x))$ then each $a \in M$ is put into at least one set Q_i. This set is unique because $\mathfrak{B}_0 \models \forall x(\mathrm{P}_i(x) \to \neg\mathrm{P}_j(x))$ if $i \neq j$. Two connected elements are in different sets of the partition because $\forall x \forall y(\mathrm{P}_i(x) \,\&\, \mathrm{P}_i(y) \to \neg\mathrm{R}(x,y))$ is valid in $\mathfrak{B}_0$. Hence $\langle M,R\rangle$ is k-colourable. □

This is the final example we shall give of the use of the compactness theorem in this section. However it is such a fundamental tool for model theory that it will be used explicitly or implicitly in much of what follows.

<u>Exercise 3.5.</u> Show that there is no finite set Σ of sentences in a first order language L_E such that $\mathfrak{A} \models \Sigma$ if and only if A is infinite.

<u>Exercise 3.6.</u> Suppose T_1,T_2 are theories in a language L such that for any realization $\mathfrak{M}$ of L, $\mathfrak{M} \models T_1$ if and only if $\mathfrak{M} \not\models T_2$. Show that T_1 and T_2 are both finitely axiomatizable. (We recall, from theorem 3.24 that T is finitely axiomatizable iff there is some finite $\Sigma \subseteq Form(L)$ such that $\Sigma \vdash \sigma$ for each $\sigma \in T$.)

<u>Exercise 3.7.</u> (<u>Herbrand's theorem</u>) Suppose $\exists\vec{x}\phi(\vec{x})$ is a universally valid existential sentence in a language L with at least one constant. Show that there is a universally valid sentence of the form $\phi(\vec{t}_1) \vee \ldots \vee \phi(\vec{t}_n)$ where for $i=1,\ldots,n$ $\vec{t}_i$ is a sequence of closed terms of L.

<u>Exercise 3.8.</u> Show that if S is a finite set and P a partial ordering on S then there is a total ordering P^* on S extending P. Hence show, using the compactness theorem that any partially ordered set $\langle X,\leq\rangle$ can be totally ordered by $\leq^*$ extending $\leq$.

<u>Exercise 3.9.</u> Show that there is no set of sentences Σ of a first order language L_E with, besides equality, one binary predicate letter P such that for any realization $\mathfrak{A}$ of L $\mathfrak{A} \models \Sigma$ if and only if P, the interpretation of P well orders A.

<u>Exercise 3.10.</u> Show that the theory of torsion-free Abelian groups is not finitely axiomatizable.

(The group $\langle G, +, 0\rangle$ is *torsion free* if for each $n \in \omega$ and each $x \in G$ such that $x \neq 0$ $\underbrace{x + x + \ldots + x}_{n \text{ times}} \neq 0$ i.e. no element other than 0 has finite order).

<u>Exercise 3.11.</u> Suppose F is the theory of fields in a language L with 0,1 , +, -, $\cdot$. Let $L^* = L \cup \{P\}$ where P is a unary predicate. OF is the theory generated by F together with

1. $\forall x(x = 0 \vee P(x) \vee P(-x))$
2. $\forall x \forall y(P(x) \mathbin{\&} P(y) \rightarrow P(x+y))$
3. $\forall x(P(x) \rightarrow \neg P(-x))$.

If $x < y$ is defined to be $P(y-x)$ then prove in OF that $<$ is a strict total ordering.

Show that there exist non-Archimedian models of OF i.e. models in which there are elements a,b such that $a,b \in P$ and for no natural number n does $n.a > b$ hold.

<u>Exercise 3.12.</u> Let X be the class of all realizations for the first order language L. For each $\sigma \in Sent(L)$ let $U_\sigma = \{\mathfrak{M}: \mathfrak{M} \in X \mathbin{\&} \mathfrak{M} \models \sigma\}$ and let $\mathcal{U}$ be the topology on X generated by the basis $\{U_\sigma : \sigma \in Sent(L)\}$. Show that the compactness theorem for L is equivalent to the assertion: $\langle X, \mathcal{U}\rangle$ is a compact space.

3.6. COMPLETENESS FOR THE PROPOSITIONAL CALCULUS

The Henkin-style proof of the completeness theorem given in sections 3.2 and 3.4. can be adapted to establish the completeness of an axiomatization of the propositional calculus. The proof in section 3.2 assumes that each instance of a tautology is provable. The full generality of this property of our particular axiom system for the predicate calculus is not used. Very few provable tautological forms are quoted in the proof. It is sufficient to assume that these particular tautologies are provable in the system. Likewise, in order to prove the completeness of a system for the propositional calculus using the Henkin method it is sufficient to establish that a few selected tautologies are theorems.

For the remainder of this section L_0 will denote a propositional language with propositional variables $\{P_i : i \in I\}$ and connectives $\neg$, &. Formulae and the remaining propositional connectives $\vee$, $\rightarrow$, $\leftrightarrow$ are defined as before (with a few simplifications) (cf. page 18). A *truth assignment* for L_0 is a map $v : Form(L_0) \rightarrow \{T, F\}$ such that

$$v(\phi \ \& \ \psi) = T \text{ if and only if } v(\phi) = T \text{ and } v(\psi) = T$$

and $\quad v(\neg\phi) = T$ if and only if $v(\phi) = F$

$\phi \in Form(L_0)$ is a *tautology* if and only if $v(\phi) = T$ for all truth assignments v.

A_0 is an axiomatization for the classical propositional calculus with the following properties:

1) all axioms are tautologies;

2) if ϕ is inferred from $\phi_1, \ldots, \phi_n$ by one of the rules of A_0 and v is some truth assignment such that $v(\phi)_i = T$ for $i = 1, \ldots, n$ then $v(\phi) = T$;

3) modus ponens is a (derived) rule of A_0 and the deduction theorem is a metatheorem;

4) the following tautologies are provable in A_0:

(a) $(\neg\phi \rightarrow \psi) \rightarrow ((\neg\phi \rightarrow \neg\psi) \rightarrow \phi)$

(b) $\phi \rightarrow (\psi \rightarrow (\phi \ \& \ \psi))$

(c) $\phi \ \& \ \psi \rightarrow \phi, \ \phi \ \& \ \psi \rightarrow \psi$

for any $\phi, \psi \in Form(L_0)$.

(The system we have given clearly has these properties.)

$\Sigma \subseteq Form(L_0)$ is *consistent* (relative to A_0) if there is no $\phi \in Form(L_0)$ such that $\Sigma \vdash_{A_0} \phi$ and $\Sigma \vdash_{A_0} \neg\phi$ where $\vdash_{A_0}$ denotes a derivation from the axioms and rules of A_0. $\Sigma \subseteq Form(L_0)$ is *complete* if for each $\phi \in Form(L_0)$ either $\Sigma \vdash_{A_0} \phi$ or $\Sigma \vdash_{A_0} \neg\phi$.

$\Sigma \subseteq Form(L_0)$ has a *model* if there is some truth assignment v such that $v(\phi) = T$ for all $\phi \in \Sigma$.

Theorem 3.28. *Suppose* $\Sigma \subseteq Form(L_0)$. Σ *is consistent if and only if* Σ *has a model.*

Proof: If $\Sigma \vdash_{A_0} \phi$ then it is straightforward to show by induction on the length of the derivation of ϕ from Σ in A_0 that a model for Σ is also a model for ϕ. This is an immediate consequence of properties 1) and 2) of A_0. A corollary of this analogue of theorem 2.6 is that a set of sentences with a model is consistent.

To show that a consistent set of sentences has a model we modify the construction in section 2. The steps A and C of the proof are not necessary. The construction of a complete consistent set Σ^* extending the consistent set Σ is almost exactly as in theorems 3.7 (in the countable case) and 3.19 (in the uncountable case). A sequence of sets of formulae $\{\Sigma_\alpha : \alpha < \beta\}$ is defined such that for each α, Σ_α is consistent. $\Sigma_0 = \Sigma$, $\Sigma_{\alpha+1}$ is Σ_α unless it is not the case that $\Sigma_\alpha \vdash \phi_\alpha$ in which case $\Sigma_{\alpha+1}$ is $\Sigma_\alpha \cup \{\neg\phi_\alpha\}$, and Σ_λ for limit λ is $\bigcup_{\alpha<\lambda} \Sigma_\alpha$. Then using the deduction theorem (property 3) and 4a we obtain a contradiction from assuming Σ_α consistent and $\Sigma_{\alpha+1}$ inconsistent. As before Σ_β is $\bigcup_{\alpha<\beta} \Sigma_\alpha$ and is complete and consistent. For a complete consistent set $\Sigma \subseteq Form(L_0)$ a canonical truth assignment v is defined as follows:

(i) $v(P_i) = T$ if and only if $\Sigma \vdash_{A_0} P_i$,

(ii) v is extended to $Form(L_0)$ in the usual way, i.e. $v(\phi \;\&\; \psi) = T$ if and only if $v(\phi) = T$ and

$v(\psi) = T.$

$v(\neg\phi) = T$ if and only if $v(\phi) = F$.

Then by induction on the length of $\phi \in Form(L_0)$ we show

$$v(\phi) = T \text{ if and only if } \Sigma \vdash_{A_0} \phi.$$

For propositional variables this is merely a restatement of the definition of v. If ϕ is $\neg\psi$ then the induction hypothesis implies that $v(\psi) = T$ if and only if $\Sigma \vdash \psi$. By consistency if $\Sigma \vdash \psi$ then not $\Sigma \vdash \phi$ and by completeness if not $\Sigma \vdash \phi$ then $\Sigma \vdash \neg\phi$. Hence $v(\psi) = F$ iff $\Sigma \vdash \phi$ and so $v(\phi) = T$ iff $\Sigma \vdash \phi$. If ϕ is ϕ_1 & ϕ_2 then by the induction hypothesis $v(\phi_1) = T$ iff $\Sigma \vdash \phi_1$ and $v(\phi_2) = T$ iff $\Sigma \vdash \phi_2$. By the assumptions 4) (b) and (c) on A_0 we have $\Sigma \vdash \phi_1$ and $\Sigma \vdash \phi_2$ just in case $\Sigma \vdash \phi_1$ & ϕ_2. So $v(\phi) = T$ iff $\Sigma \vdash \phi$ which completes the proof. □

Corollary 3.29.

(a) *A_0 is complete i.e. the theorems of A_0 are precisely the tautologies.*

(b) *If $\Sigma \subseteq Form(L_0)$ then Σ has a model if and only if every finite subset of Σ has a model.*

Proof: (a) Properties 1 and 2 are sufficient to ensure that all theorems are tautologies. Conversely if not $\vdash \phi$ then as in the proof above $\{\neg\phi\}$ is consistent and hence has a model. So all tautologies are theorems.

(b) is proved exactly as in 3.14. □

4
BEGINNING MODEL THEORY

4.1. THE LÖWENHEIM-SKOLEM THEOREMS

The Löwenheim-Skolem theorems are, like the compactness theorem, model-theoretic results which can be deduced as straightforward consequences of the completeness theorem. Both the Löwenheim-Skolem theorems concern the possible size of models of satisfiable sets of sentences. So for this section it is assumed that the language L_E has equality and that all models are normal. The first theorem concerns the possible lower bounds on the cardinality of models for satisfiable sets of sentences.

<u>Theorem 4.1. (AC)</u> <u>Downward Löwenheim-Skolem Theorem</u>
If $\Sigma \subseteq Sent(L_E)$, Σ *has cardinality* κ *and* Σ *has a model then* Σ *has a model of cardinality* $\leq \kappa$. *(It is assumed that* κ *is infinite.)*

Proof: If Σ has a model then Σ is consistent. Σ has infinite cardinality κ just in case the language of Σ has cardinality κ. Hence by theorem 3.20 Σ has a model of cardinality $\leq \kappa$.□

Löwenheim (1915) first proved a particular instance of theorem 4.1. He showed that when Σ is a finite set of sentences (and hence the language of Σ is countable) and Σ is satisfiable then Σ is satisfiable in a countable domain. The result was generalized by Skolem in 1920 to the case where Σ is a countably infinite set of formulae. The first Skolem proof assumed the axiom of choice (as indeed does the proof of the more general form of the theorem given in 4.1 above). However the completeness theorem for countable sets of sentences (and, more especially, theorem 3.16B) does not require the axiom of choice. Hence the following special case of theorem 4.1 may be proved without reference to the axiom of choice.

Theorem 4.2. *If Σ is a countable satisfiable set of sentences then Σ has a countable model.*

In 1922 Skolem published a new proof of theorem 4.2 which did not depend on the axiom of choice.

One of the paradoxes of logic, the so-called Skolem paradox, is a consequence of theorem 4.2. Using a countable language it is possible to formalize a strong-enough theory T for a system including the real numbers, in which the proposition that the reals are uncountable is derivable. The language of T contains the two predicates = and ϵ, and sets and functions are represented in the theory as elements with special defining properties. Assuming that T has a model (or, equivalently, is consistent) then theorem 4.2 implies that T has a countable model $\mathfrak{S}$. There is no inconsistency here for the following reason: the proposition that the reals are uncountable is formalized by 'There is no function f: $R \to N$ which is one-one'. In the model $\mathfrak{S}$ there are elements of the domain representing the real numbers, the natural numbers, and functions. The statement that the reals are uncountable will be true in the model; but that simply means that no element *in the model* satisfies the formula: 'x is a one-one function from R to N'. The element $R^{\mathfrak{S}}$ in the domain representing the reals may indeed be countable; no contradiction occurs because the one-one function from $R^{\mathfrak{S}}$ onto the natural numbers will be outside the model.

The compactness theorem and the downward Löwenheim-Skolem theorem are both used to prove the so called upward Löwenheim-Skolem theorem which is a non-trivial analogue of lemma 2.24 for languages with equality. It establishes that sets of sentences with an infinite model have models of arbitrarily high cardinality.

Theorem 4.3 (AC) Upward Löwenheim-Skolem Theorem. *Suppose $\Sigma \subseteq Sent(L_E)$ where L_E has cardinality κ. If Σ has an infinite model then Σ has a model of cardinality λ for* *each infinite* $\lambda \geq \kappa$.

Proof: The argument is similar to that used in theorem 3.22. Suppose λ is a some cardinal $\geq \kappa$. We extend L_E to a language L_E^* with cardinality λ by adjoining a set of new constants $\{b_i : i \in I\}$ where I is some set with cardinal λ. $\Sigma^* \subseteq Sent(L_E^*)$ is defined to be the union of Σ with the set of all sentences of the form $\neg b_i = b_j$ where $i,j \in I$ and $i \neq j$. Consider a finite subset Σ_0 of Σ^*. Only finitely many of the new constants $b_{i_1},\ldots,b_{i_k}$ occur in Σ_0. By hypothesis Σ has an infinite model $\mathfrak{A}$. $\mathfrak{A}$ can be expanded to a realization for the language of Σ^*, by adjoining as distinguished elements distinct $a_{i_1},\ldots,a_{i_k} \in A$ for $b_{i_1},\ldots,b_{i_k}$ and an arbitrary $a \in A$, say a_i, for the other b_i's. Then $\langle \mathfrak{A}, \{a_i\}_{i\in I}\rangle \models \Sigma_0$. By compactness Σ^* has a model

$\mathfrak{B} = \langle \mathfrak{B}_0, \{b_i\}_{i\in I}\rangle$ where $\mathfrak{B}_0$ is a realization for L_E. Since

$b_i \neq b_j$ if $i,j \in I$ and $i \neq j$ then $\mathfrak{B}$ has cardinality $\geq \lambda$. By the downward Löwenheim-Skolem theorem Σ^* (with cardinality λ) has a model of cardinality $\leq \lambda$, say $\langle \mathfrak{C}, \{c_i\}_{i\in I}\rangle$. But the form of Σ^* implies that any model has cardinality at least λ and hence $\mathfrak{C}$ has cardinality exactly λ. $\mathfrak{C} \models \Sigma$ and so the proof is complete. □

Theorem 4.3 was proved in the countable case by Skolem (1920) and by Tarski in the uncountable case. Just as we may infer from theorem 3.22 that no set of sentences of a first order language is valid in precisely the finite realizations of L_E so we may infer from the upward Löwenheim-Skolem theorem that there is no set of sentences characterizing the realizations of a given infinite cardinality.

The generalized Löwenheim-Skolem theorems (4.1 and 4.3) are proved by appealing to the generalized completeness theorem and its corollary, the compactness theorem, the proofs of which depend upon the axiom of choice. This dependence is intrinsic; either of the Löwenheim-Skolem theorems is sufficiently strong to imply the axiom of choice.[†]

4.2. COMPLETENESS AND CATEGORICITY.

The completeness theorem shows that the relations $\models$ and $\vdash$ are interchangeable. So syntactic properties of theories (or sets of sentences) can be expressed also in terms of semantic notions and vice versa. As has already been demonstrated in the compactness theorem and the Lowenheim-Skolem theorems, model-theoretic properties are sometimes most easily proved by considering their proof-theoretic analogues. A further illustration of the kind of flexibility in the metatheory implied by the completeness theorem is given by an alternative model-theoretic characterization of a complete theory. A preliminary definition is needed.

Definition 4.4. Suppose $\mathfrak{A}$ and $\mathfrak{B}$ are two realizations for a first order language L. $\mathfrak{A}$ is *elementarily equivalent* to $\mathfrak{B}$ (written $\mathfrak{A} \equiv \mathfrak{B}$) if and only if for each sentence $\sigma \in Sent(L)$ $\mathfrak{A} \models \sigma$ implies $\mathfrak{B} \models \sigma$.

Elementary equivalence is an equivalence relation on the class of structures for L since for any $\sigma \in sent(L)$ and any realization $\mathfrak{A}$ for L either $\mathfrak{A} \models \sigma$ or $\mathfrak{A} \models \neg\sigma$. So $\mathfrak{A} \equiv \mathfrak{B}$ iff for each $\sigma \in Sent(L)$ $\mathfrak{A} \models \sigma$ just in case $\mathfrak{B} \models \sigma$. An example of two elementarily equivalent structures was demonstrated in theorem 3.25. The two structures $\mathfrak{N}$ (the natural numbers) and the non-isomorphic $\mathfrak{M}_0$ are both models for all the true sentences in first order arithmetic. More trivial examples are provided by any pair $\mathfrak{A}, \mathfrak{B}$ of isomorphic structures.

Exercise 4.1. Suppose L is a countable first order language, and $\mathfrak{C}$ some realization of L of cardinality $\kappa \geq \omega$. Show that for each infinite λ there is a realization $\mathfrak{D}$ of L of cardinality $\geq \lambda$ which is elementarily equivalent to $\mathfrak{C}$.

[†] See e.g. Bell and Slomson (1969), pages 83-4.

<u>Lemma 4.5.</u> *$\Sigma \subseteq Sent(L)$ is complete if and only if all models of Σ are elementarily equivalent.*

Proof: If Σ is complete then (definition 3.1) for every $\sigma \in Sent(L)$ either $\Sigma \vdash \sigma$ or $\Sigma \vdash \neg\sigma$. So by the completeness theorem either $\Sigma \models \sigma$ or $\Sigma \models \neg\sigma$ and hence for all models $\mathfrak{S}$ of Σ $\mathfrak{S} \models \sigma$ or for all models $\mathfrak{S}$, $\mathfrak{S} \models \neg\sigma$. So any two models of Σ are elementarily equivalent.

Conversely suppose all models of Σ are elementarily equivalent. Then for all $\sigma \in Sent(L)$ either $\Sigma \models \sigma$ or $\Sigma \models \neg\sigma$ which by the completeness theorem implies that either $\Sigma \vdash \sigma$ or $\Sigma \vdash \neg\sigma$. □

Complete theories are of interest for several reasons. Mathematical applications can be made using the characterization given in lemma 4.5. Suppose $\mathfrak{S}$ and $\mathfrak{T}$ are two models for a complete theory T. If σ is a sentence in the language of T which is true in $\mathfrak{S}$ then σ is true in $\mathfrak{T}$. As a result of this observation known properties of a familiar structure, $\mathfrak{S}$ say, may be inferred about a less familiar structure $\mathfrak{T}$ which is a model for the same complete theory T. Another important application of the concept is concerned with the decidability of a theory. This aspect is considered in more detail in section 4.4.

Some presentations of a theory T are trivially complete. For example, if $\mathfrak{M}$ is a realization for L then

$$Th(\mathfrak{M}) = \{\sigma : \sigma \in Sent(L) \text{ and } \mathfrak{M} \models \sigma\}$$

is complete because each sentence of L is either true or false in $\mathfrak{M}$. However in the more usual mathematical situation a finite or recursive set of axioms Σ is given for a theory T (i.e. $T = \{\sigma \in Sent(L) : \Sigma \vdash \sigma\}$) and the question is then posed: is T complete? For example, P, Peano's axioms for arithmetic written in a first order language, are true in the intended interpretation, the natural numbers. Can all sentences in the first order language for arithmetic be decided in P? or, equivalently, does P generate the theory

$Th(\mathfrak{N})$? The answer to these questions is no (Gödel 1931). But now we consider how we might establish for a given theory T that T is complete. A direct attempt to show that, given any sentence σ (in the language of T), either σ or $\neg\sigma$ is a theorem of T is unlikely to be successful for it is not possible to consider all conceivable theorems of T in a finite time. An alternative model-theoretic approach, directly using the concept of elementary equivalence, is again unsatisfactory because the infinite set of all sentences of the language must be considered. The example in theorem 3.25 illustrates that elementarily equivalent structures are not necessarily 'alike' in the algebraic sense; that is, they are not necessarily isomorphic (although, trivially, isomorphic structures are elementarily equivalent). For the remainder of this chapter we will consider alternative methods by which we may decide for a given theory T if T is complete. Lemma 4.5 will be used to establish tests for completeness and mathematical consequences of a known complete theory T, rather than as a tool with which to tackle directly the problem of the completeness of a particular theory. Examples of elementarily equivalent structures will follow immediately once we have established that certain theories are complete.

Since isomorphic structures are elementarily equivalent one way to establish that a given theory T is complete is to show that all models of T are isomorphic. However a corollary of the upward Löwenheim-Skolem theorem is that any theory T (in a language with equality) with infinite models has non-isomorphic models because models of different cardinality are necessarily non-isomorphic. For languages without equality a consistent theory always has non-isomorphic models by lemma 2.24. Nearly all the familiar mathematical theories and in particular all examples to be considered here, are expressed in languages with equality. So for the remainder of this chapter the first order language L will have equality. Prompted by these considerations we introduce two further definitions related to the algebraic structure of models of a particular theory.

Definition 4.6. Suppose T is a consistent theory in a language L.

(a) T is *categorical* if all models are isomorphic.

(b) T is κ-*categorical* (where κ is a cardinal) if

(i) T has a model of cardinal κ

and (ii) any two models of T of cardinal κ are isomorphic.

From the remarks preceding the definition it is clear that any categorical theory can have only finite models. Such theories do indeed exist.

Example 4.7. The theory T_n defined by the sentence asserting the existence of precisely n elements (exercise 2.16) is trivially categorical because any two sets with n elements are isomorphic as sets.

Example 4.8. A less trivial example is the theory of a group of order 4 in which each non-identity element has order 2. This theory is axiomatized by the following:

$$\forall x \forall y \forall z(((x\cdot y)\cdot z) = (x\cdot(y\cdot z)))$$

$$\forall x(x\cdot 1 = x)$$

$$\forall x \exists y(x\cdot y = 1)$$

$$\forall x(x\cdot x = 1)$$

$$\exists x \exists y \exists z \exists w(\neg x = y \;\&\; \neg x = z \;\&\; \neg x = w \;\&\; \neg y = z$$
$$\&\; \neg y = w \;\&\; \neg z = w \;\&$$
$$\forall u(u = x \vee u = y \vee u = z \vee u = w)).$$

We have already noted that isomorphic structures are necessarily elementarily equivalent, and hence it is trivial that categorical theories are complete. The following lemma provides a partial converse.

Lemma 4.9. *A complete theory $T \subseteq Sent(L)$ with a finite model is categorical.*

Proof: Suppose $\mathfrak{A}$ is the given finite model of T. If A has n elements then $\mathfrak{A} \models \sigma_n$, where σ_n is the sentence of L asserting that there are precisely n elements (cf. exercise 2.16). Since T is complete $\sigma_n \in T$. So if $\mathfrak{B}$ is any other model of T then $\mathfrak{B}$ will also contain n elements. By lemma 4.5 $\mathfrak{A} \equiv \mathfrak{B}$. Now we must show that $\mathfrak{A}$ is actually isomorphic to $\mathfrak{B}$. This we achieve by considering a finite sequence of alphabetic extensions of L, say $\langle L_r = L \cup \{c_1,\ldots,c_r\} : r=1,\ldots,n\rangle$ (where the c_i are new constant symbols not occurring in L) and then, supposing $A = \{a_1,\ldots,a_n\}$ we show by induction on $r \le n$ that, for some (suitably chosen) numbering $\{b_1,\ldots,b_n\}$ of B, we have

$$(*)\quad \langle \mathfrak{A},a_1,\ldots,a_r\rangle \equiv \langle \mathfrak{B},b_1,\ldots,b_r\rangle \quad \text{(as structures for } L_r).$$

Finally we construct an isomorphism h which sends a_i to b_i.

First we consider the induction on r to establish (*). When $r = 0$ there is nothing to prove since we already know that $\mathfrak{A} \equiv \mathfrak{B}$. So now we assume

$$\langle \mathfrak{A},a_1,\ldots,a_r\rangle \equiv \langle \mathfrak{B},b_1,\ldots,b_r\rangle \quad \text{for some } r < n$$
$$\text{and } L_{r+1} = L \cup \{c_1,\ldots,c_r,c_{r+1}\}.$$

Suppose it is *not* the case that

$\langle \mathfrak{A},a_1,\ldots,a_{r+1}\rangle \equiv \langle \mathfrak{B},b_1,\ldots,b_r,b\rangle$ for any $b \in B-\{b_1,\ldots,b_r\}$.

Then for each $b \in B-\{b_1,\ldots,b_r\}$ there is some sentence $\phi_b(c_1,\ldots,c_r,c_{r+1})$ of L_{r+1} with
$\langle \mathfrak{A},a_1,\ldots,a_{r+1}\rangle \models \phi_b(c_1,\ldots,c_r,c_{r+1})$ and
$\langle \mathfrak{B},b_1,\ldots,b\rangle \models \neg\phi_b(c_1,\ldots,c_{r+1})$.

Now we consider the formula *in* L_r that is the (finite) conjunction of

$$\{\phi_b(c_1,\ldots,c_r,x) : b \in B-\{b_1,\ldots,b_r\}\} \cup \{\neg c_i = x : i=1,\ldots,r\}$$

which we denote by $\theta(c_1,\ldots,c_r,x)$.

Now $\langle \mathfrak{A},a_1,\ldots,a_r\rangle \models \theta(c_1,\ldots,c_r,x)\ [a_{r+1}]$
and $\langle \mathfrak{B},b_1,\ldots,b_r\rangle \not\models \theta(c_1,\ldots,c_r,x)\ [b]$

for any $b \in B-\{b_1,\ldots,b_r\}$.

So $\langle\mathfrak{A},a_1,\ldots,a_r\rangle \models \exists x\theta(c_1,\ldots,c_r,x)$. But then by our induction hypothesis $\langle\mathfrak{B},b_1,\ldots,b_r\rangle \models \exists x\theta(c_1,\ldots,c_r,x)$. This is a contradiction since if $\langle\mathfrak{B},b_1,\ldots,b_r\rangle \models \theta(c_1,\ldots,c_r,x)\,[b]$ then necessarily $b \neq b_i$ for $i=1,\ldots,r$ and so $b \in B-\{b_1,\ldots,b_r\}$. So there is some $b \in B-\{b_1,\ldots,b_r\}$, which we will call b_{r+1}, such that

$$\langle\mathfrak{A},a_1,\ldots,a_{r+1}\rangle \equiv \langle\mathfrak{B},b_1,\ldots,b_{r+1}\rangle .$$

Now we can conclude that

$$\langle\mathfrak{A},a_1,\ldots,a_n\rangle \equiv \langle\mathfrak{B},b_1,\ldots,b_n\rangle$$

and consider the map $h\colon A \to B$ defined by $h(a_i) = b_i$ for $i=1,\ldots,n$.

If R is a k-ary predicate letter of L then for any $a_{i_1},\ldots,a_{i_k} \in A$

$$\mathfrak{A} \models R[a_{i_1},\ldots,a_{i_k}] \text{ iff } \langle\mathfrak{A},a_1,\ldots,a_n\rangle \models R(c_{i_1},\ldots,c_{i_k}).$$

By the elementary equivalence of $\langle\mathfrak{A},a_1,\ldots,a_n\rangle$ and $\langle\mathfrak{B},b_1,\ldots,b_n\rangle$

$$\langle\mathfrak{A},a_1,\ldots,a_n\rangle \models R(c_{i_1},\ldots,c_{i_k}) \text{ iff } \langle\mathfrak{B},b_1,\ldots,b_n\rangle \models R(c_{i_1},\ldots,c_{i_k}).$$

So $\mathfrak{A} \models R[a_{i_1},\ldots,a_{i_k}]$ iff $\mathfrak{B} \models R[b_{i_1},\ldots,b_{i_k}]$ or, equivalently, $\langle a_{i_1},\ldots,a_{i_k}\rangle \in R^{\mathfrak{A}}$ iff $\langle h(a_{i_1}),\ldots,h(a_{i_k})\rangle \in R^{\mathfrak{B}}$.

Similarly by considering the sentences of L_n

$f(c_{i_1},\ldots,c_{i_j}) = c_p$, $c = c_i$ (f, c are respectively function and constant symbols of L)

we can show that

$$f^{\mathfrak{A}}(a_{i_1},\ldots,a_{i_j}) = a_p \text{ iff } f^{\mathfrak{B}}(h(a_{i_1}),\ldots,h(a_{i_j})) = h(a_p)$$

and $h(c^{\mathfrak{A}}) = c^{\mathfrak{B}}$.[†]

So by exercise 1.7. $\mathfrak{A} \cong \mathfrak{B}$. □

Remark: At first sight it seems plausible that the use of a well-ordering of the domain of $\mathfrak{A}$ and transfinite induction might enable us to extend the proof of lemma 4.9 to the case where A is infinite. However we know that the Löwenheim-Skolem theorems exclude this possibility. It is instructive to observe just where the proof breaks down. Certainly if $\mathfrak{A}, \mathfrak{B} \models T$ and T is complete, $\mathfrak{A}$ infinite then $\mathfrak{B}$ is also infinite ($\mathfrak{A}$ is a model for: 'there are at least n elements'). But consider the definition of the formula $\theta(c_1,\ldots,c_r,x)$. For this to be a formula it must be of finite length and so the set $B-\{b_1,\ldots,b_r\}$ must be finite for our construction to work. The essentially finite character of the expressions of a first order language is what prevents us from generalizing lemma 4.9.

The technique introduced in lemma 4.9 of adjoining to the language L for the structure $\mathfrak{A}$, constants for the elements of A does have applications, however, in the case where A is infinite. Some of these will be considered in the next section.

Exercise 4.2. Show that if Σ is categorical and the language of Σ contains only finitely many non-logical symbols then there is a finite $\Sigma_0 \subseteq \Sigma$ such that $\Sigma_0 \vdash \sigma$ for each $\sigma \in \Sigma$.

It is clear that a theory with both finite and infinite models is not complete for the sentence 'there are exactly n elements' is undecided for some finite n. We now turn our

---

[†] When we consider several realizations $\mathfrak{U},\mathfrak{B},\mathfrak{W}$ etc. for the language L we use the notation $R^{\mathfrak{U}},R^{\mathfrak{B}},R^{\mathfrak{W}}$ etc. for the interpretation of the predicate R in $\mathfrak{U},\mathfrak{B},\mathfrak{W}$ etc. Similar notations are used for function and constant symbols.

attention to theories all of whose models are infinite. Many mathematical structures of interest are infinite and hence any (partial) axiomatization of the theory of such a structure will not be categorical by the Löwenheim-Skolem theorem. However there are examples of such axiomatizations that are κ-categorical for some cardinals κ.

Example 4.10. The sentence which asserts that there are at least n elements: $\exists x_1 \ldots\ldots \exists x_n(\underset{1 \le i < j \le n}{\&} (\neg x_i = x_j))$ is κ-categorical for each $\kappa \ge n$.

Example 4.11. Suppose L is language with a single binary predicate $<$ (besides equality). We write $x < y$ for $<(x,y)$. DLO (dense linear (or total) ordering without end points) is the theory with the following axioms:

(i) $\forall x(\neg x < x)$
(ii) $\forall x \forall y(x < y \vee y < x \vee x = y)$
(iii) $\forall x \forall y \forall z(x < y \;\&\; y < z \rightarrow x < z)$
(iv) $\forall x \forall y(x < y \rightarrow \exists z(x < z \;\&\; z < y)$
(v) $\forall x \exists y(y < x)$
(vi) $\forall x \exists y(x < y)$.

(i)-(iii) are the axioms for a strict linear ordering. (iv) asserts that the ordering is dense, i.e. between any two distinct elements lies a third. (v), (vi) assert that there are no least or greatest elements in the ordering. Any model of DLO is necessarily infinite since a linear ordering of a finite set cannot be dense, and moreover always has a least element. Two non-isomorphic models of DLO are $\langle Q,<\rangle$, the ordered set of rationals, and $\langle R,<\rangle$, the ordered set of reals. A theorem of Cantor establishes that DLO is $\aleph_0$-categorical.

Theorem 4.12. (Cantor 1895). *If* $\langle A,<_A\rangle$ *and* $\langle B,<_B\rangle$ *are two countable models of DLO then* $\langle A,<_A\rangle \cong \langle B,<_B\rangle$.

Proof: We assume that

$$A = \{a_0, a_1, \ldots\} \text{ and } B = \{b_0, b_1, \ldots\}$$

We do *not* assume that the order of enumeration of the domain of either model bears any relation to the interpretation of $<$ in that model. To show that $\mathfrak{A} = \langle A, <_A\rangle$ and $\mathfrak{B} = \langle B, <_B\rangle$ are isomorphic we exhibit a one-one function h from A onto B which is order-preserving i.e. $h(a) <_B h(a')$ iff $a <_A a'$ for all $a, a' \in A$. We begin by setting $h(a_0) = b_0$. At the $2n$th stage in the construction of h we ensure that $h(a_0), \ldots, h(a_n)$ have been defined and at the $(2n+1)$th stage that $b_0, \ldots, b_n$ are in the range. Suppose that at the mth stage h has been defined on $\{a_{i_1}, \ldots, a_{i_k}\}$ and for $j=1,\ldots,k$ $h(a_{i_j}) = b_{r_j}$; also that for $1 \le j, j' \le k$ $a_{i_j} <_A a_{i_{j'}}$ iff $b_{r_j} <_B b_{r_{j'}}$.

Case 1. If $m = 2n$ then there are two possibilities.

1.1. If $a_n \in \{a_{i_1}, \ldots, a_{i_k}\}$ then do nothing.

1.2. If $a_n \notin \{a_{i_1}, \ldots, a_{i_k}\}$ then consider the order relations holding between a_n and a_{i_j} for $j=1,\ldots,k$. Since B is densely ordered by $<_B$ and has no first or last element there is a uniquely defined minimal p such that for $j=1,\ldots,k$ $b_p <_B b_{r_j}$ iff $a_n <_A a_{i_j}$. There are in fact infinitely many elements of B which are in the same order relation to $b_{r_1}, \ldots, b_{r_k}$ as a_n to $a_{i_1}, \ldots, a_{i_k}$ but there is a first such b in the particular (well ordered) listing we have adopted for elements of B. Let $h(a_n) = b_p$.

Case 2. If $m = 2n+1$ then again there are two cases.

2.1. If $b_n \in \{b_{r_1}, \ldots, b_{r_k}\}$ then do nothing.

2.2. If $b_n \notin \{b_{r_1}, \ldots, b_{r_k}\}$ then there is a minimal q such that $a_q <_A a_{i_j}$ iff $b_n <_B b_{r_j}$ for $j=1,\ldots,k$. Let $h(a_q) = b_n$.

Because $<_A, <_B$ are strict orderings on A, B respectively it is immediate from the definition that h is one-one. h is defined on the whole of A and is onto B since by the $(2k+1)$'th stage $h(a_k)$ has been defined and b_k is in the range. By construction h is order-preserving and consequently h is an isomorphism with the requisite properties. □

Categoricity in a countable power does not imply κ-categoricity for uncountable κ as is demonstrated in the following lemma.

Lemma 4.13. *Suppose* $\langle R, <\rangle$ *is the ordered set of reals and* $\langle R^*, <\rangle$ *is the substructure of* $\langle R, <\rangle$ *obtained by deleting* 0 *from the domain. Then* $\langle R, <\rangle$ *is not isomorphic to* $\langle R^*, <\rangle$.

Proof: Suppose $\langle R,<\rangle \cong \langle R^*, <\rangle$ and $h\colon \mathfrak{R} \to \mathfrak{R}^*$ is an isomorphism. For each non zero n, $\frac{1}{n} \in R^*$. Let p_n be the inverse image under h of $\frac{1}{n}$, i.e. $h(p_n) = \frac{1}{n}$. Then since h is order-preserving $p_n < p_{n-1} < \ldots < p_1$. The strictly decreasing sequence $\{p_n\}_{n\in\omega}$ is bounded below (by, for example, q where $h(q) = -1$). So, since R is complete, $\{p_n\}$ converges to some $p \in R$. $h(p) < h(p_n)$ for each n. Hence $h(p) < 0$. But then there is some $r \in R^*$ such that $h(p) < r < 0$ and $r < h(p_n) = \frac{1}{n}$ for each n. So since p is the greatest lower bound of $\{p_n\}_{n\in\omega}$ $h^{-1}(r) \le p$. This is a contradiction since h is order preserving. Consequently such an isomorphism h cannot exist. □

Corollary 4.14. *DLO is not* $2^{\aleph_0}$*-categorical.*

Proof: $\langle R,<\rangle$ and $\langle R^*,<\rangle$ are both models of DLO of cardinal $2^{\aleph_0}$. □

Exercise 4.3. Let L be the language for DLO, $\mathfrak{D}$ a realization of L, $Th(\mathfrak{D}) = \{\sigma : \sigma \in Sent(L)\ \&\ \mathfrak{D} \models \sigma\}$. Give finite axiomatizations for the following theories and show that they are $\aleph_0$-categorical:

(i) $Th(\langle [0,1], \lessdot \rangle)$
(ii) $Th(\langle [0,1), \lessdot \rangle)$
(iii) $Th(\langle (0,1], < \rangle)$.

Exercise 4.4. (i) Suppose $\langle X, \lessdot \rangle$ is a countable linearly ordered set. Show that $\langle X, \lessdot \rangle$ can be embedded in $\langle Q, < \rangle$. When is $\langle X, < \rangle$ elementarily equivalent to $\langle Q, < \rangle$?

(ii) Use the compactness theorem to show that any linearly ordered set can be embedded in a dense linear ordering without end points.

The following example demonstrates that precisely the opposite situation to that for DLO can occur. There is a consistent theory which is $2^{\aleph_0}$-categorical but is not $\aleph_0$-categorical.

Example 4.15. A recursive set of axioms for fields of characteristic zero, F_0, was given in theorem 3.24. Suppose ACF_0 is $F_0 \cup \{P_n\}_{\substack{n\in\omega \\ n\neq 0}}$ where P_n is the sentence asserting that every non-constant polynomial of degree n has a root; i.e. P_n is

$$\forall y_n \forall y_{n-1} \ldots \forall y_0 (\neg y_n = 0 \rightarrow \exists x(y_n x^n + y_{n-1} x^{n-1} + \ldots + y_0 = 0))$$

where x^m is an abbreviation for $\underbrace{(\ldots (x \cdot x) \cdot x) \ldots x)}_{\text{m times}}$.

ACF_0 is a set of axioms for algebraically closed fields of characteristic zero. The field of algebraic numbers, A, is a countable model for ACF_0 and the field of complex numbers an uncountable model. A theorem of Steinitz (1910) asserts that ACF_0 is λ-categorical for uncountable λ. However ACF_0 is not $\aleph_0$-categorical. Both A and the algebraic closure of the field extension of A obtained by adjoining one transcendental element are countable models for ACF_0.

Exercise 4.5. Suppose T_S is the theory (in the language L with a unary function S and a constant 0) generated by the following axioms:

(i) $\forall x(\neg 0 = Sx)$;

(ii) $\forall x \forall y(Sx = Sy \to x = y)$;

(iii) $\forall x(\neg x = 0 \to \exists y(x = Sy))$;

(iv) $\forall x(\neg x = Sx)$, $\forall x(\neg x = S(Sx))$, $\forall x(\neg x = S(S(S(x)))$, $\ldots$, $\forall x(\neg x = S(S\ldots(Sx)\ldots))$, $\ldots$

(a) Show that all models of T_S are infinite and that $\langle N, S, O\rangle$ is a model for T_S.

(b) Find another non-isomorphic countable model for T_S.

(c) Show that if $\mathfrak{S} \models T_S$ and $\mathfrak{S}$ has uncountable cardinality κ then $\mathfrak{S}$ is isomorphic to the disjoint union of $\langle N, S, O\rangle$ with κ copies of $\langle Z, S\rangle$ (the integers together with the successor function); deduce that T_S is κ-categorical.

We have exhibited examples of the following types of theory T:

(i) T is κ-categorical for each κ (example 4.10);

(ii) T is $\aleph_0$-categorical but not $2^{\aleph_0}$-categorical (theorem 4.12 and exercise 4.3);

(iii) T is λ-categorical for uncountable λ but not $\aleph_0$-categorical (example 4.15 and exercise 4.5).

The following important theorem is due to Morley (but will not be proved here).

Result 4.16. (Morley 1965) *If T is a countable theory and T is λ-categorical for some uncountable λ then T is λ-categorical for each uncountable λ.*

Having established that κ-categoricity is a meaningful concept we return to the starting point for the discussion of categoricity to consider how it relates to completeness.

Theorem 4.17. (Vaught 1954) *Suppose T is a theory of car-*

dinality κ *with no finite models. If* T *is* λ-*categorical for some (infinite)* $\lambda \geq \kappa$ *then* T *is complete.*

Proof: Suppose T is not complete but is λ-categorical for some $\lambda \geq \kappa$. Then there is some sentence σ in the language of T such that neither $T \vdash \sigma$ nor $T \vdash \neg\sigma$. So by lemma 3.6 $T_1 = T \cup \{\neg\sigma\}$ and $T_2 = T \cup \{\sigma\}$ are both consistent. By the completeness theorem, for $i=1,2$ T_i has a model $\mathfrak{M}_i$ which by hypothesis is not finite. Then by the upward Löwenheim-Skolem theorem (4.3) T_i has a model $\mathfrak{N}_i$ of cardinal λ. $\mathfrak{N}_1$ is a model for $\neg\sigma$ and $\mathfrak{N}_2$ for σ. However we are assuming that T is λ-categorical so $\mathfrak{N}_1$ and $\mathfrak{N}_2$ are isomorphic structures. Since isomorphic structures satisfy precisely the same sentences this is a contradiction. Hence if T is λ-categorical then T is complete. □

Corollary 4.18 (a) *DLO,* ACF_0 *and* T_S *(exercise 4.5) are complete.*

(b) $\langle Q, \lessdot \rangle \equiv \langle R, \lessdot \rangle$ *and the field of algebraic numbers is elementarily equivalent to the field of complex numbers.*

Proof: (a) Theorem 4.17 is immediately applicable: *DLO*, ACF_0 and T_S are countable theories with no finite models; *DLO* is $\aleph_0$-categorical by Cantor's theorem (4.12) and ACF_0, T_S are κ-categorical for each uncountable κ.

(b) $\langle Q, \lessdot \rangle$ and $\langle R, \lessdot \rangle$ are both models for *DLO*. The field of algebraic numbers and the field of complex numbers are models for ACF_0. The result now follows from lemma 4.5. □

This corollary demonstrates that Vaught's test (4.17) is a useful technique with which to tackle the problem of completeness. However there are complete theories which are not κ-categorical for any infinite κ and for such theories we must develop alternative methods for establishing completeness. We mention here just one example of such a theory, the theory of discretely ordered sets with first but no last element. We denote this theory by T_{DIS}. It has the following axioms:

(i) $\forall x(\neg x < x)$
(ii) $\forall x \forall y \forall z(x < y \ \& \ y < z \rightarrow x < z)$ (< is a total ordering)
(iii) $\forall x \forall y(x < y \vee y < x \vee y = x)$
(iv) $\forall x(\exists y(y < x) \rightarrow \exists z(z < x \ \& \ \forall w(w < x \rightarrow (w < z \vee w = z))))$

(This says: every element with predecessors has an immediate predecessor.)

(v) $\forall x(\exists y(x<y) \rightarrow \exists z(x<z \ \& \ \forall w(x<w \rightarrow (z < w \vee z = w))))$

(Every element with successors has an immediate successor.)

(vi) $\exists x \forall y(x < y \vee x = y)$ (There is a first element.)
(vii) $\forall x \exists y(x < y)$ (There is no last element.)

T_{DIS} is not $\aleph_0$-categorical because, for example, linearly ordered sets of type ω and of type $\omega + \omega^* + \omega$[†] are countable non-isomorphic models of the theory. T_{DIS} is not $2^{\aleph_0}$-categorical either: if λ denotes the (natural) order type of the reals then ordered sets with order types $\omega + (\omega^* + \omega)\lambda$ and $\omega + (\omega^* + \omega)(\lambda+1)$ provide non isomorphic models of cardinality $2^{\aleph_0}$. By Morley's result we infer that T_{DIS} is not κ-categorical for any cardinal κ. We cannot, therefore, use Vaught's test here. In the following section an alternative test for completeness will be used to show that T_{DIS} is nevertheless complete.

4.3. ELEMENTARY EMBEDDINGS AND MODEL COMPLETENESS

It follows from the results of the previous section that complete theories exist having non-isomorphic models. Elementary equivalence is therefore a coarser relation than isomorphism on the class of structures for a given language (that is, there are fewer equivalence classes). The following definition introduces a new binary relation on the class of L-structures which is strong enough to imply elementary equivalence and yet does not imply isomorphism.

[†]see e.g. Cantor (1895) for the definitions of addition and multiplication of order types.

Definition 4.19. If $\mathfrak{A} \subseteq \mathfrak{B}$ then $\mathfrak{A}$ is an *elementary substructure* of $\mathfrak{B}$ if for all $\phi(\vec{x}) \in Form(L)$ and $\vec{a} \in A$, $\mathfrak{A} \models \phi[\vec{a}]$ implies $\mathfrak{B} \models \phi[\vec{a}]$. We write $\mathfrak{A} \prec \mathfrak{B}$ and say $\mathfrak{B}$ is an *elementary extension* of $\mathfrak{A}$.

We recall that for $\mathfrak{A}$ and $\mathfrak{B}$ to be elementarily equivalent (definition 4.4) we stipulated that the same *sentences* be true in both structures. The relation we are considering here refers to all *formulae*. If $\mathfrak{A} \subseteq \mathfrak{B}$ then for $\mathfrak{A}$ to be an elementary substructure of $\mathfrak{B}$ not only must the same sentences be true in $\mathfrak{A}$ and $\mathfrak{B}$ but also a formula satisfiable in $\mathfrak{A}$ must be satisfied in $\mathfrak{B}$ by the *same* elements of A.

Exercise 4.6. Show that '... $\mathfrak{A} \models \phi[\vec{a}]$ implies $\mathfrak{B} \models \phi[\vec{a}]$...' in definition 4.19 may be replaced by '..$\mathfrak{A} \models \phi[\vec{a}]$ iff $\mathfrak{B} \models \phi[\vec{a}]$...'.

Exercise 4.7. Show that $\prec$ is a partial ordering on the class of L-structures.

Exercise 4.8. Suppose $\mathfrak{C}, \mathfrak{D}, \mathfrak{E}$ are structures of the same type such that $\mathfrak{C} \prec \mathfrak{E}, \mathfrak{D} \prec \mathfrak{E}$ and $\mathfrak{C} \subseteq \mathfrak{D}$. Show that $\mathfrak{C} \prec \mathfrak{D}$.

Definition 4.20. The embedding $f: \mathfrak{A} \to \mathfrak{B}$ is an *elementary embedding* if $f(\mathfrak{A}) \prec \mathfrak{B}$. We write $f: \mathfrak{A} \overset{\prec}{\to} \mathfrak{B}$. When such an embedding exists $\mathfrak{A}$ is said to be *elementarily embeddable* in $\mathfrak{B}$.

An embedding from $\mathfrak{A}$ to $\mathfrak{B}$ preserves the algebraic structure of $\mathfrak{A}$ so that the image of $\mathfrak{A}$ is an isomorphic copy of $\mathfrak{A}$ inside $\mathfrak{B}$. An elementary embedding preserves much more as is demonstrated in the following lemma.

Lemma 4.21. *Suppose $\mathfrak{A}$, $\mathfrak{B}$ are structures of the same type and $f: A \to B$ is any map.*

(a) *f is an embedding iff for every atomic formula $\phi(\vec{x})$ in L and $\vec{a} \in A$, $\mathfrak{A} \models \phi[\vec{a}]$ implies $\mathfrak{B} \models \phi[f\vec{a}]$;*

(b) *f is an elementary embedding iff for every formula* $\phi(\vec{x})$ *in L and* $\vec{a} \in A$, $\mathfrak{A} \models \phi[\vec{a}]$ *implies* $\mathfrak{B} \models \phi[f\vec{a}]$.

Proof: The proof of (a) is left as an exercise. To prove (b) we suppose firstly that f is an elementary embedding according to definition 4.20. Suppose $\phi(\vec{x}) \in Form(L)$ and $\vec{a} \in A$: then since $\mathfrak{A}$, $f(\mathfrak{A})$ are isomorphic $\mathfrak{A} \models \phi[\vec{a}]$ iff $f(\mathfrak{A}) \models \phi[f\vec{a}]$ (exercise 1.15). So if $\mathfrak{A} \models \phi[\vec{a}]$ then, since $f(\mathfrak{A}) \prec \mathfrak{B}$ and $f\vec{a} \in f(A)$ $\mathfrak{B} \models \phi[f\vec{a}]$.

Conversely suppose $f: A \to B$ is a map such that for every formula $\phi(\vec{x})$ or L and $\vec{a} \in A$, $\mathfrak{A} \models \phi[\vec{a}]$ implies $\mathfrak{B} \models \phi[f\vec{a}]$. We must show firstly that f is an isomorphism between $\mathfrak{A}$ and its image, $f(\mathfrak{A})$ (definition 1.9). Suppose $t(x_1,\ldots,x_n) \in Term(L)$ and $\phi(y,x_1,\ldots,x_n)$ is the formula

$$t(x_1,\ldots,x_n) = y.$$

(We are assuming in this chapter that L is a language with equality.) Then if $t^{\mathfrak{A}}[a_1,\ldots,a_n]$ is b we have

$$\mathfrak{A} \models \phi[b,a_1,\ldots,a_n]$$

and hence $\mathfrak{B} \models \phi[fb,fa\ ,\ldots,fa_n]$ or, equivalently

$$t\ [fa_1,\ldots,fa_n] = fb.$$

Hence if g is a function symbol in L with interpretations $g^{\mathfrak{A}}$ and $g^{\mathfrak{B}}$ in $\mathfrak{A}$ and $\mathfrak{B}$ respectively then

$$f(g^{\mathfrak{A}}(a_1,\ldots,a_n)) = g^{\mathfrak{B}}(fa_1,\ldots,fa_n)$$

(take $t(x_1,\ldots,x_n)$ to be $g(x_1,\ldots,x_n)$), and if c is a constant symbol with interpretations $c^{\mathfrak{A}}$, $c^{\mathfrak{B}}$ then $f(c^{\mathfrak{A}}) = c^{\mathfrak{B}}$ (take t to be c). Furthermore f is one-one (take the formula $\neg x = y$).

Now suppose $\langle a_1,\ldots,a_{\lambda(i)}\rangle \in R_i^{\mathfrak{A}}$. Then if $\phi(x_1,\ldots,x_{\lambda(i)})$ is the formula $R(x_1,\ldots,x_{\lambda(i)})$ we have $\mathfrak{A} \models \phi[a_1,\ldots,a_{\lambda(i)}]$ and hence $\mathfrak{B} \models \phi[fa_1,\ldots,fa_{\lambda(i)}]$, or, equivalently, $\langle fa_1,\ldots,fa_{\lambda(i)}\rangle \in R_i^{\mathfrak{B}}$. Conversely if $\langle a_1,\ldots,a_{\lambda(i)}\rangle \notin R_i^{\mathfrak{A}}$

then consideration of the formula $\neg R_i(x_1,\ldots,x_{\lambda(i)})$ leads us to conclude that $\langle fa_1,\ldots,fa_{\lambda(i)}\rangle \notin R_i^{\mathfrak{B}}$. So by exercise 1.7 we conclude that f is an isomorphism between $\mathfrak{A}$ and $f(\mathfrak{A})$.

Now let $\phi(x_1,\ldots,x_n)$ be any formula of L, $a_1,\ldots,a_n \in A$: If $f(\mathfrak{A}) \models \phi[fa_1,\ldots,fa_n]$ then $\mathfrak{A} \models \phi[a_1,\ldots,a_n]$ whence by our hypothesis, $\mathfrak{B} \models \phi[fa_1,\ldots,fa_n]$. So $f(\mathfrak{A}) \prec \mathfrak{B}$ which completes the proof.□

It is a trivial consequence of the definitions that if $\mathfrak{U}$ is an elementary substructure of $\mathfrak{B}$, or more generally, if $\mathfrak{U}$ is elementarily embeddable in $\mathfrak{B}$ then $\mathfrak{U} \equiv \mathfrak{B}$. If $\mathfrak{U} \equiv \mathfrak{B}$ then it is certainly not the case in general that $\mathfrak{U} \prec \mathfrak{B}$ or even that $\mathfrak{U}$ is elementarily embeddable in $\mathfrak{B}$. Even if $\mathfrak{U} \subseteq \mathfrak{B}$ and $\mathfrak{U} \equiv \mathfrak{B}$ still no converse to the above result holds; each of the following three possibilities can be realized:

(i) $\mathfrak{U} \prec \mathfrak{B}$;

(ii) there is an f such that $f:\mathfrak{U} \overset{\prec}{\to} \mathfrak{B}$ although $\mathfrak{U} \not\prec \mathfrak{B}$;

(iii) there is no f such that $f:\mathfrak{U} \overset{\prec}{\to} \mathfrak{B}$.

Illustrations of these situations are given below.

<u>Example 4.22.</u> $\langle Q,<\rangle \prec \langle R,<\rangle$. By corollary 4.18 (b) we already know that $\langle Q,<\rangle \equiv \langle R,<\rangle$. Trivially $\langle Q,<\rangle \subseteq \langle R,<\rangle$. Suppose that $\phi(\vec{x},y) \in Form(L)$ and for some $\vec{q} \in Q$, $r \in R$ $\mathfrak{R} \models \phi[\vec{q},r]$. We show that then

(*) there is some $q_0 \in Q$ such that $\mathfrak{R} \models \phi[\vec{q},q_0]$.

The rationals in the sequence $\vec{q}$ form a finite linearly ordered set which without loss of generality we may take as $q_1 < q_2 < \ldots < q_n$. Then either $r \in Q$ (when there is nothing to prove) or $r < q_1$ or $r > q_n$ or for some $i=1,\ldots,n-1$ $q_i < r < q_{i+1}$. In each of the last three cases it is possible to define an order isomorphism $h:\mathfrak{R} \to \mathfrak{R}$ such that $h(q_j) = q_j$ for $j=1,\ldots,n$ and $h(r) \in Q$. The last case is illustrated diagrammatically:

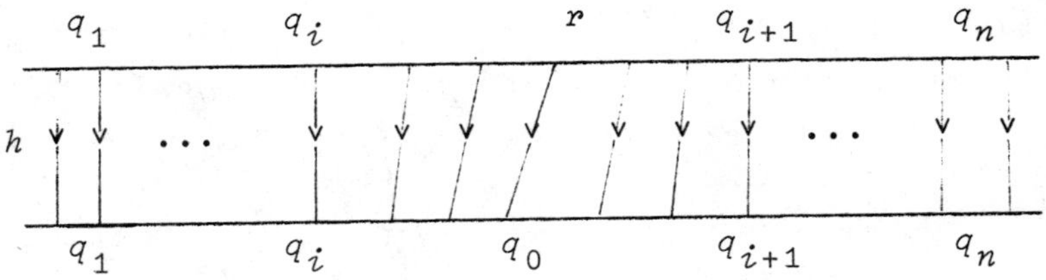

More precisely: $h(x) = x$ for $x \leq q_i$ or $x \geq q_{i+1}$; let q_0 be some rational such that $q_i < q_0 < q_{i+1}$ and then define

$h(x) = q_i + (x-q_i)\dfrac{(q_0-q_i)}{(r - q_i)}$ for $q_i < x \leq r$;

$h(x) = q_{i+1} - (q_{i+1}-x)\dfrac{(q_{i+1} - q_0)}{(q_{i+1} - r)}$ for $r < x < q_{i+1}$.

(*) follows immediately when $h(r)$ is taken for q_0. Now it is straightforward to show by induction on the length of ϕ that for $\vec{q} \in Q$

$$\langle Q, \lessdot\rangle \models \phi[\vec{q}] \quad \text{implies} \quad \langle R, \lessdot\rangle \models \phi[\vec{q}].$$

(*) is used for the induction step when ϕ is of the form $\forall x\,\psi(x)$.

<u>Example 4.23.</u> Suppose $R_0 = R \cap [0,\infty)$ and $R_1 = R \cap [1,\infty)$. Then $\langle R_1, \lessdot\rangle \subseteq \langle R_0, \lessdot\rangle$ but it is not the case that $\langle R_1, \lessdot\rangle \prec \langle R_0, \lessdot\rangle$, for suppose $\phi(x)$ is $\exists y(y < x)$, then

$$\langle R_1, \lessdot\rangle \models \neg\phi[1] \quad \text{and} \quad \langle R_0, \lessdot\rangle \models \phi[1].$$

Moreover $\langle R_0, \lessdot\rangle \equiv \langle R_1, \lessdot\rangle$ since both are models for the theory of dense linear orderings with first but no last element which is complete by exercise 4.3 (ii) and Vaught's test.

If f: $R_1 \to R_0$ is defined by $f(x) = x-1$ then f is an elementary embedding from $\langle R_1, \lessdot\rangle$ to $\langle R_0, \lessdot\rangle$ (it is actually an isomorphism).

<u>Example 4.24.</u> Suppose A is the disjoint union of $\aleph_0$-copies of the positive rationals and $\aleph_1$ copies of the non-negative

rationals; B is the disjoint union of $\aleph_0$ copies of the non-negative rationals and $\aleph_1$ copies of the rationals. Specifically:

$$A = \{(\alpha,q) : (0 \leq \alpha < \omega \ \& \ q > 0 \ \& \ q \in Q) \vee (\omega \leq \alpha < \omega_1 \ \& \ q \geq 0 \ \& \ q \in Q)\}$$

$$B = \{(\alpha,q) : (0 \leq \alpha < \omega \ \& \ q \geq 0 \ \& \ q \in Q) \vee (\omega \leq \alpha < \omega_1 \ \& \ q \in Q)\}.^{\dagger}$$

For both A and B a relation $<$ is defined:

$$(\alpha_1,q_1) < (\alpha_2,q_2) \quad \text{iff} \quad \alpha_1 = \alpha_2 \text{ and } q_1 <_Q q_2$$

Now take $\mathfrak{A} = \langle A,<\rangle$ and $\mathfrak{B} = \langle B,<\rangle$ and certainly $\mathfrak{A} \subseteq \mathfrak{B}$. To show that $\mathfrak{A}$ and $\mathfrak{B}$ are elementarily equivalent we construct an axiomatizable theory T, say, for which both $\mathfrak{A}$ and $\mathfrak{B}$ are models and then prove that the theory T is complete. T has the following axioms:

(i) $\forall x(\neg x < x)$, } '< is a strict partial order'

(ii) $\forall x \forall y \forall z(x < y \ \& \ y < z \rightarrow x < z)$,

(iii) $\forall x \forall y(x < y \rightarrow \exists z(x < z \ \& \ z < y))$, '< is dense'

(iv) $\forall x \exists y(x < y)$ 'there are no maximal elements'

(v) $\exists x_1 \ldots . \exists x_n(\underset{1\leq i<j\leq n}{\&}(\neg x_i = x_j) \ \& \ \underset{1\leq i\leq n}{\&} \forall y(\neg y < x_i))$ for each n

'there are (at least) n distinct minimal elements'.

(vi) $\exists x_1 \ldots . \exists x_n(\underset{1\leq i<j\leq n}{\&}(\neg x_i = x_j) \ \& \ \underset{1\leq i,j\leq n}{\&}(\neg x_i < x_j) \ \& \ \underset{1\leq i\leq n}{\&} \forall y(y < x_i \vee y = x_i \rightarrow \exists z(z < y)))$

for each n,

'there are (at least) n incomparable elements $x_1,\ldots,x_n$ with no minimal element $y \leq x_i$ for $i=1,\ldots,n$.'

$^{\dagger}\omega_1$ is the second infinite initial ordinal and has cardinal $\aleph_1$.

It is clear that $\mathfrak{A}$ and $\mathfrak{B}$ are models for this theory. We leave it as an exercise to show that T is complete. (Hint: show that T is $\aleph_0$-categorical and apply Vaught's test.) Now we show, by contradiction, that there is no elementary embedding from $\mathfrak{A}$ into $\mathfrak{B}$; for suppose $f:\mathfrak{A}\to\mathfrak{B}$ is an elementary embedding. From considerations of cardinality for some $\alpha \geq \omega$ $f(\alpha,0) = (\beta,q)$ for some $\beta \geq \omega$. Suppose ϕ is the formula $\exists x(x < y)$. Then

$$\mathfrak{A} \models \neg\phi[(\alpha,0)] \text{ but } \mathfrak{B} \models \phi[f(\alpha,0)].$$

Hence f cannot be elementary.

The concept of elementary embedding can be reduced to that of elementary equivalence by considering appropriate expansions of the original structures. In Chapter 3 alphabetic extensions of a given first order language L were constructed to provide instantiating constants for certain derivable sentences of the form $\exists x\phi(x)$. Model-theoretic considerations lead to the following general construction of an alphabetic extension of a given language. (This construction generalizes that of L_r in lemma 4.9.)

Definition 4.25. If $\mathfrak{A}$ is a realization for L, $L^+(\mathfrak{A})$ is the alphabetic extension of L obtained by adjoining to L a new individual constant a for each a in A.

Even when the language L already contains a constant c that is interpreted as a in $\mathfrak{A}$ an additional constant a is included in $L^+(\mathfrak{A})$. The expanded structure for $L^+(\mathfrak{A})$ in which each new constant has its intended interpretation (as the element to which it corresponds) is denoted by $\langle\mathfrak{A}, A\rangle$.

Lemma 4.26. *If* $f:A \to B$ *then* $f:\mathfrak{A} \overset{\prec}{\to} \mathfrak{B}$ *iff* $\langle\mathfrak{A},A\rangle \equiv \langle\mathfrak{B},f(A)\rangle$

Proof: With each $\phi \in Form\, L^+(\mathfrak{A})$ we associate a $\phi_0 \in Form(L)$ as follows: suppose $\mathsf{a}_1,\ldots,\mathsf{a}_n$ are the constants of $L^+(\mathfrak{A})$ (and not in L) occurring in ϕ; $\phi_0(y_1,\ldots,y_n)$ is the formula obtained when each constant a_i is replaced by a distinct new variable y_i not already occuring in ϕ. Now suppose that $\langle \mathfrak{C}, A \rangle$ is some realization for the language $L^+(\mathfrak{A})$. We show by induction on the length of $\phi(\vec{x}) \in Form\ L^+(\mathfrak{A})$ that

(*) $\langle \mathfrak{C}, A \rangle \models \phi(a_1, \ldots, a_n)[\vec{c}]$ iff $\mathfrak{C} \models \phi_0[a_1, \ldots, a_n, \vec{c}]$.

(*) is trivial to verify when ϕ is atomic or is of the form $\neg\psi$, ϕ_1 & ϕ_2. So we turn immediately to the case where ϕ is $\forall x\psi(x)$. By the induction hypothesis for each $c_0 \in C$

$$\langle \mathfrak{C}, A \rangle \models \psi(a_1, \ldots, a_n)[c_0, \vec{c}\,] \text{ iff } \mathfrak{C} \models \psi_0[a_1, \ldots, a_n, c_0, \vec{c}].$$

So $\langle \mathfrak{C}, A \rangle \models \forall x\psi(a_1, \ldots, a_n, x)[\vec{c}]$ iff $\mathfrak{C} \models \forall x\psi_0[a_1, \ldots, a_n, \vec{c}]$ or since ϕ_0 is $\forall x\psi_0(y_1, \ldots, y_n, x)$,

$$\langle \mathfrak{C}, A \rangle \models \phi(a_1, \ldots, a_n)[\vec{c}] \text{ iff } \mathfrak{C} \models \phi_0[a_1, \ldots, a_n, \vec{c}].$$

Now suppose $f: \mathfrak{A} \overset{\prec}{\to} \mathfrak{B}$. Then for $\phi \in Sent\ L^+(\mathfrak{A})$ (and so $\phi_0 \in Form(L)$)

$$\mathfrak{A} \models \phi_0[a_1, \ldots, a_n] \text{ iff } \mathfrak{B} \models \phi_0[fa_1, \ldots, fa_n]$$

and then by (*)

$$\langle \mathfrak{A}, A \rangle \models \phi \text{ iff } \langle \mathfrak{B}, f(A) \rangle \models \phi$$

which implies that $\langle \mathfrak{A}, A \rangle \equiv \langle \mathfrak{B}, f(A) \rangle$. The converse follows in a similar manner. □

<u>Corollary 4.27.</u> *If* $\mathfrak{A} \subseteq \mathfrak{B}$ *then* $\mathfrak{A}$ *is an elementary substructure of* $\mathfrak{B}$ *iff* $\langle \mathfrak{A}, A \rangle \equiv \langle \mathfrak{B}, A \rangle$.

We have already introduced existential and universal formulae (exercise 1.22). A further classification of formulae is helpful at this stage.

<u>Definition 4.28.</u> Suppose $\phi \in Form(L)$ and $n \geq 0$.
$\phi \in \forall n\ Form(L)$ iff there is a prenex normal form ϕ' for ϕ such that ϕ' has the form

$$\forall \vec{x}_1 \exists \vec{x}_2 \forall \vec{x}_3 \ldots Q_n \vec{x}_n \psi$$

where Q_n is $\forall$ if n is odd, $\exists$ if n is even and ψ is quantifier-free.

$\phi \in \exists n\ Form(L)$ iff $\neg\phi \in \forall n\ Form(L)$

$\forall n\ Sent(L) = \forall n\ Form(L) \cap Sent(L)$ and $\exists n\ Sent(L)$ is defined analogously.

The important characteristic in this classification of (prenex) formulae is not the number of quantifiers in the formula but rather the number of *alternations* of quantifiers. Some examples will help to clarify this idea. We suppose that ψ is a quantifier-free formula:

$$\forall x_1 \forall x_2 \exists y_1 \psi \in \forall 2\ Form(L)$$

$$\forall x_1 \forall x_2 \forall x_3 \psi \in \forall 1\ Form(L)$$

$$\forall x_1 \exists y_1 \forall x_2 \psi \in \forall 3\ Form(L)$$

$$\exists y_1 \forall x_1 \forall x_2 \psi \in \exists 2\ Form(L).$$

For an arbitrary formula ϕ the associated prenex normal form ϕ_0 is not unique and so, according to our definition, ϕ may belong to more than one of the sets $\forall n\ Form(L)$ and $\exists n\ Form(L)$. For example, suppose ϕ is

$$\exists y \psi_1(y) \to \exists x \psi_2(x)$$

where ψ_1 and ψ_2 are quantifier-free. Then both

$$\forall z\ \exists w(\psi_1(z) \to \psi_2(w))$$

and

$$\exists w\ \forall z(\psi_1(z) \to \psi_2(w))$$

(where z does not occur in ψ_2, w does not occur in ψ_1) are prenex normal forms for ϕ. So $\phi \in \forall 2\ Form(L) \cap \exists 2\ Form(L)$.

As trivial corollaries to the definition we remark that the set of quantifier-free formulae is $\forall 0\ Form(L)$ (which is the same as $\exists 0\ Form(L)$); the set of universal formulae is contained in $\forall 1\ Form(L)$; the set of existential formulae is contained in $\exists 1$ *Form*(L); and finally,

$$Form(L) = \bigcup_{n\in\omega} \forall n\ Form(L) \quad \text{(since trivially } \exists n\ Form(L) \subseteq \forall n{+}1\ Form(L);\ \vdash \exists x\phi(x) \leftrightarrow \forall y \exists x\phi(x) \text{ if } y \text{ does not occur in } \exists x\phi(x)).$$

A weakening of the conditions for an elementary substructure leads to the following definition.

Definition 4.29. If $\mathfrak{A} \subseteq \mathfrak{B}$ then $\mathfrak{A}$ is an *n-elementary substructure* of $\mathfrak{B}$ if for all $\phi(\vec{x}) \in \forall n\ Form(L)$ and $\vec{a} \in A$ $\mathfrak{A} \models \phi[\vec{a}]$ implies $\mathfrak{B} \models \phi[\vec{a}]$. We write $\mathfrak{A} \prec_n \mathfrak{B}$

Exercise 4.9. Show that $\mathfrak{A} \prec_n \mathfrak{B}$ iff $\mathfrak{A} \subseteq \mathfrak{B}$ and for all $\phi(\vec{x}) \in \exists n\ Form(L)$ and $\vec{a} \in A$

$$\mathfrak{A} \models \phi[\vec{a}] \text{ implies } \mathfrak{B} \models \phi[\vec{a}].$$

Exercise 4.10. (a) Show that $\mathfrak{A} \subseteq \mathfrak{B}$ iff $\mathfrak{A} \prec_0 \mathfrak{B}$
(b) Show that $\mathfrak{A} \prec \mathfrak{B}$ iff $\mathfrak{A} \prec_n \mathfrak{B}$ for all n.

Further characterizations of the relations $\subseteq, \prec$ are obtained when diagrams, a notion first introduced by Robinson, are considered.

Definition 4.30. $\mathfrak{A}$ is a realization for L.

(a) The *open diagram* of $\mathfrak{A}$, $\Delta(\mathfrak{A}) \subseteq Sent\ L^{+}(\mathfrak{A})$ is $\{\phi : \langle \mathfrak{A}, A\rangle \models \phi$ and ϕ is an atomic sentence or the negation of an atomic sentence in $L^{+}(\mathfrak{A})\}$.

(b) The *complete diagram* of $\mathfrak{A}$, $\Delta^{c}(\mathfrak{A})$ is $Th(\mathfrak{A}, A) = \{\sigma : \langle \mathfrak{A}, A\rangle \models \sigma$ and $\sigma \in Sent\ L^{+}(\mathfrak{A})\}$.

The open diagram is an encoding in the expanded language $L^{+}(\mathfrak{A})$ of the structure of $\mathfrak{A}$: for every $\langle a_1, \ldots, a_{\lambda(i)}\rangle \in R_i$ the sentence $\mathrm{R}_i(\mathrm{a}_1, \ldots, \mathrm{a}_{\lambda(i)}) \in \Delta(\mathfrak{A})$ and conversely if $\langle a_1, \ldots, a_{\lambda(i)}\rangle \notin R_i$ then $\neg\mathrm{R}_i(\mathrm{a}_1, \ldots, \mathrm{a}_{\lambda(i)}) \in \Delta(\mathfrak{A})$; similarly, $f_j(a_1, \ldots, a_{\mu(j)}) = a$ iff $\mathrm{f}_j(\mathrm{a}_1, \ldots, \mathrm{a}_{\mu(j)}) = \mathrm{a} \in \Delta(\mathfrak{A})$.
For example, when $\mathfrak{A}$ is the field of integers modulo p,

$\mathfrak{J}_p = \langle J_p, +, \cdot, 0, 1\rangle$ then $\Delta(\mathfrak{A})$ essentially gives the multiplication tables for + and $\cdot$. Trivially $\Delta(\mathfrak{A}) \subseteq \Delta^c(\mathfrak{A})$ and $\Delta^c(\mathfrak{A})$, being the set of all true sentences in a structure, is complete. The following exercise demonstrates the close connection between the notions we have introduced in this section.

Exercise 4.11. Show that: (a) there is an elementary embedding $f: \mathfrak{A} \overset{\prec}{\to} \mathfrak{B}$ iff there is an $X \subseteq B$ such that $\langle \mathfrak{B}, X\rangle \models \Delta^c(\mathfrak{A})$.

(b) there is an n-elementary embedding $f: \mathfrak{A} \overset{\prec_n}{\to} \mathfrak{B}$ (i.e. $f(\mathfrak{A}) \prec_n \mathfrak{B}$) iff there is an $X \subseteq B$ such that $\langle \mathfrak{B}, X\rangle \models \Delta^c(\mathfrak{A}) \cap \forall n\ Sent\ L^+(\mathfrak{A})$.

(c) there is an embedding $f: \mathfrak{A} \to \mathfrak{B}$ iff there is an $X \subseteq B$ such that $\langle \mathfrak{B}, X\rangle \models \Delta(\mathfrak{A})$.

In lemma 4.5 we established a model-theoretic condition (in terms of elementary equivalence) for a given theory to be complete. An alternative condition in terms of elementary embeddings is given in the following lemma.

Lemma 4.31. *T is complete iff given any two models of T $\mathfrak{A}, \mathfrak{B}$ there is a structure $\mathfrak{C}$ and elementary embeddings $f: \mathfrak{A} \overset{\prec}{\to} \mathfrak{C}$, $g: \mathfrak{B} \overset{\prec}{\to} \mathfrak{C}$*

Proof: Suppose firstly that T is complete and that $\mathfrak{A}, \mathfrak{B}$ are any two models of T. By exercise 4.11(a) above for $\mathfrak{C}$ to be an elementary extension of $\mathfrak{A}$ it is necessary and sufficient that for some $X \subseteq C$ $\langle \mathfrak{C}, X\rangle \models \Delta^c(\mathfrak{A})$. So we must show that for some $\mathfrak{C}$, $X \subseteq C$, $Y \subseteq C$ we have

$$\langle \mathfrak{C}, X, Y\rangle \models \Delta^c(\mathfrak{A}) \cup \Delta^c(\mathfrak{B})$$

where X is the set of interpretations for $\{\mathrm{a}: a \in A\}$ and Y is the set of interpretations for $\{\mathrm{b}: b \in B\}$. By the completeness theorem it will be sufficient to show that $\Delta^c(\mathfrak{A}) \cup \Delta^c(\mathfrak{B})$ is consistent. We will work by contradiction. If $\Delta^c(\mathfrak{A}) \cup \Delta^c(\mathfrak{B})$ is inconsistent then some finite subset is inconsistent; so there exist sentences $\phi(\vec{\mathrm{a}}) \in Sent\ L^+(\mathfrak{A})$,

$\psi(\vec{b}) \in Sent\ L^+(\mathfrak{B})$ such that $\langle\mathfrak{A}, A\rangle \models \phi(\vec{a})$, $\langle\mathfrak{B}, B\rangle \models \psi(\vec{b})$, and $\{\phi(\vec{a}), \psi(\vec{b})\}$ is inconsistent. Hence $\phi(\vec{a}) \vdash \neg\psi(\vec{b})$. The constants $\vec{b}$ are distinct from $\vec{a}$ and so we may replace $\vec{b}$ by distinct new variables $\vec{y}$ not occurring in $\phi(\vec{a})$ or $\psi(\vec{b})$ to obtain $\phi(\vec{a}) \vdash \neg\psi(\vec{y})$. Then by generalization $\phi(\vec{a}) \vdash \forall\vec{y}\neg\psi(\vec{y})$. Since $\langle\mathfrak{A}, A\rangle \models \phi(\vec{a})$ it follows that $\langle\mathfrak{A}, A\rangle \models \forall\vec{y}\neg\psi(\vec{y})$ and hence, since $\forall\vec{y}\neg\psi(\vec{y}) \in Sent(L)$ and T is complete, $\forall\vec{y}\neg\psi(\vec{y}) \in T$. We infer that $\mathfrak{B} \models \forall\vec{y}\neg\psi(\vec{y})$. This is a contradiction since $\mathfrak{B} \models \psi[\vec{b}]$. So our assumption, that $\Delta^c(\mathfrak{A}) \cup \Delta^c(\mathfrak{B})$ is inconsistent, is false. Now by completeness there is a model $\langle\mathfrak{C}, X, Y\rangle \models \Delta^c(\mathfrak{A}) \cup \Delta^c(\mathfrak{B})$. If $X = \{x_a : a \in A\}$ and we define $f : \mathfrak{A} \to \mathfrak{C}$ by $f(a) = x_a$ then f is an elementary embedding (by exercise 4.11(a)). Similarly we define an elementary embedding $g : \mathfrak{B} \to \mathfrak{C}$ by $g(b) = y_b$ where $Y = \{y_b : b \in B\}$.

Conversely if $\mathfrak{A}, \mathfrak{B} \models T$ and there exist $f : \mathfrak{A} \preceq \mathfrak{C}$, $g : \mathfrak{B} \preceq \mathfrak{C}$ then $\mathfrak{A} \equiv \mathfrak{C} \equiv \mathfrak{B}$ and hence by lemma 4.5 T is complete. □

As with lemma 4.5 the above lemma is not in general a practical tool with which to try to establish the completeness of a given theory. However we are still concerned to establish sufficient model-theoretic conditions under which a given theory T is complete and such that the conditions are in practice sometimes verifiable. Given an arbitrary structure $\mathfrak{S}$ which is a model for T then, in general, a substructure $\mathfrak{S}'$ is unlikely to be an elementary substructure of $\mathfrak{S}$. But suppose we focus our attention on those substructures which are also models for T. We saw in example 4.22 that the structure $\langle R, <\rangle$ and its substructure $\langle Q, <\rangle$ are both models for DLO and that in this case $\langle Q, <\rangle \prec \langle R, <\rangle$. Is it possible that given any substructure $\mathfrak{E}$ of a structure $\mathfrak{D}$ such that $\mathfrak{D}, \mathfrak{E}$ are both models for DLO then $\mathfrak{E} \prec \mathfrak{D}$? The answer turns out to be yes. The situation is sufficiently general to merit the introduction of a new definition.

Definition 4.32 (Robinson). T is *model-complete* if for every pair of models of T, $\mathfrak{S}, \mathfrak{T}$ such that $\mathfrak{S} \subseteq \mathfrak{T}$, $\mathfrak{S} \prec \mathfrak{T}$.

An additional test for completeness follows.

Theorem 4.33. (Robinson's Prime Model Test 1956). *If T is model complete and there is a $\mathfrak{T} \models T$ such that $\mathfrak{T}$ can be embedded in every model for T then T is complete.*

Proof: We aim to show that, under the hypotheses of this theorem, any two models of T are elementarily equivalent so that we can apply lemma 4.5.

Suppose $\mathfrak{U}, \mathfrak{B} \models T$. By hypothesis $\mathfrak{T}$ is embeddable in $\mathfrak{U}$ and $\mathfrak{B}$ and so there are substructures $\mathfrak{U}^*, \mathfrak{B}^*$ of $\mathfrak{U}, \mathfrak{B}$ respectively such that $\mathfrak{T} \cong \mathfrak{U}^* \cong \mathfrak{B}^*$. So, since $\mathfrak{T} \models T$ then $\mathfrak{U}^*, \mathfrak{B}^* \models T$. Hence by model completeness it follows that $\mathfrak{U}^* \prec \mathfrak{U}$ and $\mathfrak{B}^* \prec \mathfrak{B}$. So $\mathfrak{U} \equiv \mathfrak{U}^* \equiv \mathfrak{B}^* \equiv \mathfrak{B}$ which completes the proof. □

Although Vaught's test was used to prove the completeness of DLO (corollary 4.18) the prime model test is also applicable in this case. Here the 'prime model' ($\mathfrak{D}$, embeddable in every model of DLO) is $\langle Q,<\rangle$. To prove that DLO is model-complete requires a new technique. In example 4.22 we showed that $\langle Q,<\rangle \prec \langle R,<\rangle$ by constructing an isomorphism from $\langle R,<\rangle$ to itself which fixed a finite number of given rationals and sent a given real to a rational. The isomorphism was defined using the properties of R as an ordered field (i.e. the operations of addition, subtraction, and division). In an arbitrary model of DLO there is no analogue. A method which can be used to establish that DLO is model-complete will be given below (corollary 4.38).

An example of the application of Robinson's prime model test to a theory for which Vaught's test cannot be used will be given after we have established some useful equivalent formulations of model-completeness. One such characterization is a seemingly weaker condition on the models of a theory T. It is an abstraction to the general case of a property of algebraically closed fields.

Definition 4.34. If $\mathfrak{A} \models T$, $\mathfrak{A}$ is *existentially closed* over the theory T if for all $\phi(\vec{x}) \in \exists 1\ Form(L)$ and $\vec{a} \in A$ and any $\mathfrak{B} \models T$ such that $\mathfrak{A} \subseteq \mathfrak{B}$ if $\mathfrak{B} \models \phi[\vec{a}]$ then $\mathfrak{A} \models \phi[\vec{a}]$.

Equivalently, $\mathfrak{A}$ is existentially closed over T if for every $\mathfrak{B}$ such that $\mathfrak{A} \subseteq \mathfrak{B} \models T$, $\mathfrak{A} \prec_1 \mathfrak{B}$.[†]

In the case where T is the theory of fields consider the existential formula $\exists y(x_n y^n + x_{n-1} y^{n-1} + \ldots + x_0 = 0)$ (where $n > 0$). If $\mathfrak{F}$ is some field, then in the usual algebraic sense $\mathfrak{F}$ is said to be algebraically closed if any non-constant polynomial $a_n y^n + \ldots + a_0$ with coefficients in F has a root in F. For any field $\mathfrak{F}$ the polynomial $a_n y^n + \ldots + a_0$ always has a root in some extension field of $\mathfrak{F}$ (e.g. the extension obtained by adjoining the root!) and hence we may say that $\mathfrak{F}$ is algebraically closed if for any $a_n, \ldots, a_0 \in F$ and any $\mathfrak{F}^* \models T$ such that $\mathfrak{F} \subseteq \mathfrak{F}^*$

if $\quad \mathfrak{F}^* \models \exists y(x_n y^n + \ldots + x_0 = 0)\,[a_n, \ldots, a_0]$ then

$$\mathfrak{F} \models \exists y(x_n y^n + \ldots + x_0 = 0)\,[a_n, \ldots, a_0].$$

We now prove a general lemma which we shall use to show the connection between model completeness and existentially closed structures over a theory T.

Lemma 4.35. *If* $\mathfrak{A} \subseteq \mathfrak{B}$ *then for* $n \geq 0$ *the following are equivalent:*

(i) $\quad \mathfrak{A} \prec_{n+1} \mathfrak{B}$

(ii) *there is some* $\mathfrak{C}$ *with* $\mathfrak{A} \prec \mathfrak{C}$, $\mathfrak{B} \prec_n \mathfrak{C}$

(iii) *there is some* $\mathfrak{C}$ *with* $\mathfrak{A} \prec_{n+1} \mathfrak{C}$, $\mathfrak{B} \prec_n \mathfrak{C}$.

Proof: That (ii) $\Rightarrow$ (iii) is an immediate corollary of exercise 4.10. (iii) $\Rightarrow$ (i) is left as an exercise (c.f. exercise 4.8).

The hardest part of the lemma is to show (i) $\Rightarrow$ (ii). So suppose $\mathfrak{A} \prec_{n+1} \mathfrak{B}$. By exercise 4.11 we must show that $\Delta^c(\mathfrak{A}) \cup [\Delta^c(\mathfrak{B}) \cap \forall n\ Sent\ L^+(\mathfrak{B})]$ (where the new constant for $a \in A \subseteq B$ is the same in both $L^+(\mathfrak{A})$ and $L^+(\mathfrak{B})$) has a model. Suppose it does not. Then by compactness there exist

[†] see Simmons (1972) for discussion of existentially closed structures.

$$\phi_1(\vec{b},\vec{a}),\ldots,\quad \phi_p(\vec{b},\vec{a}) \in \forall n\ Sent(L^+(\mathfrak{B}))$$

such that $\langle \mathfrak{B}, B\rangle \models \phi_i(\vec{b},\vec{a})$ for $i=1,\ldots,p$
and $\Delta^c(\mathfrak{A}) \cup \{\phi_1(\vec{b},\vec{a}),\ldots,\phi_p(\vec{b},\vec{a})\}$ has no model.
(Here, $\vec{b}$ is a list of the constants corresponding to $b \in B-A$ and $\vec{a}$ is a list of the constants corresponding to $a \in A$).
The sentence $\exists\vec{x} \underset{1\le i\le p}{\&} \phi_i(\vec{x},\vec{a}) \in \exists n+1\ Sent\ L^+(\mathfrak{A})$
and $\Delta^c(\mathfrak{A}) \cup \{\exists\vec{x} \underset{1\le i\le p}{\&} \phi_i(\vec{x},\vec{a})\}$ is inconsistent.
So $\langle \mathfrak{A}, A\rangle \models \forall\vec{x}\neg \underset{1\le i\le p}{\&} \phi_i(\vec{x},\vec{a})$ and
$\forall\vec{x}\neg \underset{1\le i\le p}{\&} \phi_i(\vec{x},\vec{a}) \in \forall n+1\ Sent(L^+(\mathfrak{A}))$.
By hypothesis $\mathfrak{A} \prec_{n+1} \mathfrak{B}$ and hence
$\langle \mathfrak{B}, A\rangle \models \forall\vec{x}\neg \underset{1\le i\le p}{\&} \phi_i(\vec{x},\vec{a})$ which contradicts
$\mathfrak{B} \models \phi_i[\vec{b},\vec{a}]$ for each i.

So $\Delta^c(\mathfrak{A}) \cup [\Delta^c(\mathfrak{B}) \cap \forall n\ Sent\ L^+(\mathfrak{B})]$ has a model $\langle \mathfrak{C}, B_c\rangle$ say. By a suitable renaming of the elements of B_c corresponding to the constants $\{b : b \in B\}$ we may suppose

$$\mathfrak{A} \prec \mathfrak{C} \text{ and } \mathfrak{B} \prec_n \mathfrak{C} \text{ as required.} \qquad \square$$

Now we return to the discussion of model-completeness and prove the theorem relating it to the concept of 'existentially closed'.

<u>Theorem 4.36.</u> *The following are equivalent conditions on a theory T:*

(i) *T is model-complete;*
(ii) *for every $\mathfrak{A} \models T$, $T \cup \Delta(\mathfrak{A})$ is complete;*
(iii) *all models for T are existentially closed over T.*

Proof: (i) ⇒ (ii). Suppose T is model-complete, $\mathfrak{A} \models T$ but $T \cup \Delta(\mathfrak{A})$ is not complete. Then there is some $\sigma \in Sent\ L^+(\mathfrak{A})$ such that $T \cup \Delta(\mathfrak{A}) \cup \{\sigma\}$ and $T \cup \Delta(\mathfrak{A}) \cup \{\neg\sigma\}$ are both consistent. We suppose without loss of generality that $\mathfrak{A} \models \sigma$. Because $T \cup \Delta(\mathfrak{A}) \cup \{\neg\sigma\}$ is consistent it has a model $\langle \mathfrak{B}, X \rangle$ say. By exercise 4.11 there is an embedding $f: \mathfrak{A} \to \mathfrak{B}$ and, by identifying $\mathfrak{A}$ and $f(\mathfrak{A})$ we may assume $\mathfrak{A} \subseteq \mathfrak{B}$. T is model-complete and so $\mathfrak{A} \prec \mathfrak{B}$ and, in particular, for the sentence $\neg\sigma$, since $\mathfrak{B} \models \neg\sigma$ then $\mathfrak{A} \models \neg\sigma$. This is a contradiction. So $T \cup \Delta(\mathfrak{A})$ is complete.

(ii) $\Rightarrow$ (iii). Suppose that for every model $\mathfrak{A}$ for T, $T \cup \Delta(\mathfrak{A})$ is complete. Let $\mathfrak{A}, \mathfrak{B} \models T$ such that $\mathfrak{A} \subseteq \mathfrak{B}$ and suppose $\phi(\vec{x}) \in \exists 1\ Form(L)$, $\vec{a} \in A$ such that $\mathfrak{B} \models \phi[\vec{a}]$. Then $\phi(\vec{a}) \in Sent\ L^+(\mathfrak{A})$ and hence is decided in $T \cup \Delta(\mathfrak{A})$. $\langle \mathfrak{B}, A \rangle \models T \cup \Delta(\mathfrak{A})$ by exercise 4.11 and hence $T \cup \Delta(\mathfrak{A}) \models \phi(\vec{a})$. So $\langle \mathfrak{A}, A \rangle \models \phi(\vec{a})$ and hence $\mathfrak{A} \models \phi[\vec{a}]$.

(iii) $\Rightarrow$ (i). Suppose all models for T are existentially closed over T. We show by induction on $n \geq 1$ that for all $\mathfrak{A}, \mathfrak{B}$ such that $\mathfrak{A}, \mathfrak{B} \models T$ and $\mathfrak{A} \subseteq \mathfrak{B}$, $\mathfrak{A} \prec_n \mathfrak{B}$. For $n=1$ this is simply a result of $\mathfrak{A}$ being existentially closed over T.

Now suppose the result holds for $n=k\geq 1$ and all $\mathfrak{A} \subseteq \mathfrak{B}, \mathfrak{A}, \mathfrak{B} \models T$. By the induction hypothesis $\mathfrak{A} \prec_k \mathfrak{B}$. By lemma 4.35 (i) $\Rightarrow$ (ii) there is some $\mathfrak{C}$ with $\mathfrak{A} \prec \mathfrak{C}$ and $\mathfrak{B} \prec_{k-1} \mathfrak{C}$. Since $\mathfrak{A} \prec \mathfrak{C}$ it follows that $\mathfrak{C} \models T$. So, again by the induction hypothesis since $\mathfrak{B} \subseteq \mathfrak{C}$ and $\mathfrak{B}, \mathfrak{C} \models T$, $\mathfrak{B} \prec_k \mathfrak{C}$. So appealing to the lemma again ((ii) $\Rightarrow$ (i)) we have $\mathfrak{A} \prec_{k+1} \mathfrak{B}$. Now by exercise 4.10 we conclude that if $\mathfrak{A} \subseteq \mathfrak{B}$ and $\mathfrak{A}, \mathfrak{B} \models T$ then $\mathfrak{A} \prec \mathfrak{B}$ i.e. T is model-complete. □

<u>Definition 4.37.</u> A *primitive formula* is one of the form $\exists \vec{x} \underset{1 \leq i \leq n}{\&} \phi_i(\vec{x}, \vec{y})$ where each $\phi_i(\vec{x}, \vec{y})$ is either an atomic formula or the negation of an atomic formula.

<u>Corollary 4.38.</u> (Robinson's model-completeness test)
T is model-complete iff for all $\mathfrak{A}, \mathfrak{B} \models T$ such that $\mathfrak{A} \subseteq \mathfrak{B}$ and all primitive $\phi(\vec{y})$, $\vec{a} \in A$, if $\mathfrak{B} \models \phi[\vec{a}]$ then $\mathfrak{A} \models \phi[\vec{a}]$.

Proof: First we assume the condition on the right. By

theorem 4.36 to show T model-complete it is sufficient to show that if $\mathfrak{A} \subseteq \mathfrak{B}$, $\mathfrak{A}, \mathfrak{B} \models T$ then $\mathfrak{A} \prec_1 \mathfrak{B}$. So suppose $\psi(\vec{x}) \in \exists 1\ Form(L)$ and $\mathfrak{B} \models \psi[\vec{a}]$ for some $\vec{a} \in A$.

Since a quantifier free formula is equivalent to one in disjunctive normal form[†] $\psi(\vec{x})$ is logically equivalent to a formula of the form:

$$\exists \vec{y} \bigvee_{1 \le i \le n} \phi_i(\vec{x}, \vec{y})$$

where for $i = 1, \ldots, n$ ϕ_i is a conjunction of atomic formulae and negations of atomic formulae.

So $\mathfrak{B} \models \exists \vec{y} \bigvee_{1 \le i \le n} \phi_i(\vec{x}, \vec{y})[\vec{a}]$. So for some i where $1 \le i \le n$, $\mathfrak{B} \models \exists \vec{y}\, \phi_i(\vec{x}, \vec{y})[\vec{a}]$. $\exists \vec{y}\, \phi_i(\vec{x}, \vec{y})$ is primitive. So by our hypothesis $\mathfrak{A} \models \exists \vec{y}\, \phi_i(\vec{x}, \vec{y})[\vec{a}]$ and hence $\mathfrak{A} \models \psi[\vec{a}]$.

The converse implication is trivial. □

At the end of section 4.2 we showed that T_{DIS} (the theory of discrete linear orderings with first but no last element) did not satisfy the conditions for Vaught's test. We cannot apply Robinson's prime model test directly either; T_{DIS} is not model-complete, as is readily verified by considering the structures $\langle N, < \rangle$ and $\langle N - \{0\}, < \rangle$ both of which are models for T_{DIS}, the second is a substructure of the first but $\langle N - \{0\}, < \rangle \not\prec \langle N, < \rangle$ since they have different first elements. Instead, we extend the language of T_{DIS} by adjoining a new constant 0 (for the minimal element) and a binary predicate S (for the successor relation). Now we consider the theory T^* where T^* is the theory generated by

$$T_{DIS} \cup \{\forall x (x = 0 \leftrightarrow \forall y \neg\, y < x)$$
$$\forall x \forall y (S(x, y) \leftrightarrow x < y \,\&\, \neg \exists z (x < z \,\&\, z < y))\}$$

We are going to prove that T^* is complete by applying Robinson's prime model test. It is clear that $\langle N, <, S^N, 0 \rangle$

[†]See e.g. Mendelson (1964) page 27.

will be embeddable in every model of T^*. So we must now show that T^* is model-complete. We shall use Robinson's model-completeness test. We suppose that $\mathfrak{A}, \mathfrak{B} \models T^*$, $\mathfrak{A} \subseteq \mathfrak{B}$ and $\exists \vec{x}\psi(\vec{x},\vec{y})$ is a primitive formula, $\vec{a} \in A$ such that

$$\mathfrak{B} \models \exists \vec{x}\psi(\vec{x},\vec{y})[\vec{a}]$$

So there exist elements $\vec{b} \in B$ such that $\mathfrak{B} \models \psi[\vec{b},\vec{a}]$.

The formula $\psi(\vec{x},\vec{y}) \in Form(L)$ is the conjunction of formulae of the form:

$$t_1 < t_2,\ S(t_1,t_2),\ t_1 = t_2,\ \neg t_1 = t_2,\ \neg t_1 < t_2,\ \neg S(t_1,t_2)$$

where $t_i \in \{\vec{x},\vec{y},0\}$ for $i=1,2$.

We construct a new formula $\psi^*(\vec{x},\vec{y},\vec{z}) \in Form(L)$ which is the conjunction of atomic formulae of the form

$$t_1 < t_2,\ S(t_1,t_2)$$

(we eliminate those of the remaining four forms listed above) such that

$$\mathfrak{B} \models \exists \vec{z}\psi^*(\vec{x},\vec{y},\vec{z})[\vec{b},\vec{a}] \quad \text{and for } \vec{a}^* \in A$$

$$\mathfrak{A} \models \exists \vec{z}\psi^*(\vec{x},\vec{y},\vec{z})[\vec{a}^*,\vec{a}] \text{ just in case } \mathfrak{A} \models \psi[\vec{a}^*,\vec{a}].$$

With each term t in $\{\vec{x},\vec{y},0\}$ we associate an element t^B of B as follows: (i) if t is x_i then $t^B = b_i$,

(ii) if t is y_j then $t^B = a_j$,

(iii) if t is 0 then $t^B = 0^{\mathfrak{B}}$ $(= 0^{\mathfrak{A}})$.

We are now ready to construct ψ^*.

A) Firstly we replace each conjunct of the form

$$\neg S(t_1,t_2) \text{ in } \psi.$$

Since $\mathfrak{B} \models \psi[\vec{b},\vec{a}]$ it follows that $\mathfrak{B} \models \neg S[t_1^B,t_2^B]$.

Replace $\neg S(t_1,t_2)$ by (i) $t_2 < t_1$ if $t_2^B < t_1^B$;

(ii) $t_2 = t_1$ if $t_2^B = t_1^B$;

(iii) $t_1 < z_i$ & $z_i < t_2$ if there is a $b^* \in B$ with $t_1^B < b^* < t_2^B$ and $\neg S(t_1,t_2)$ is the i'th conjunct of ψ.

B) Now we replace each conjunct $\neg t_1 < t_2$ by
(i) $t_2 < t_1$ if $t_2^B < t_1^B$;
(ii) $t_2 = t_1$ if $t_2^B = t_1^B$ ($\mathfrak{B} \models \neg t_1^B < t_2^B$).

C) Thirdly we replace each conjunct $\neg t_1 = t_2$ by
(i) $t_1 < t_2$ if $t_1^B < t_2^B$;
(ii) $t_2 < t_1$ if $t_2^B < t_1^B$.

D) The transformed formula is now the conjunction of formulae of the form: $t_1 < t_2$, $S(t_1,t_2)$, $t_1 = t_2$.
(a) If each conjunct is of the form $t_1 = t_2$ then we define ψ^* to be $0 < z_1$.
(b) Otherwise for each conjunct of the form $t_1 = t_2$ replace each occurrence of t_2 in the other conjuncts by t_1 and drop $t_1 = t_2$.

By construction it is apparent that $\psi^*(\vec{x},\vec{y},\vec{z})$ has the following properties:

(i) $\mathfrak{B} \models \exists\vec{z}\psi^*(\vec{x},\vec{y},\vec{z})[\vec{b},\vec{a}]$;
(ii) for each $\vec{a}^* \in A$

$$\mathfrak{A} \models \psi[\vec{a}^*,\vec{a}] \quad \text{iff} \quad \mathfrak{A} \models \exists\vec{z}\psi^*(\vec{x},\vec{y},\vec{z})[\vec{a}^*,\vec{a}].$$

Since we are trying to show that $\mathfrak{A} \models \psi[\vec{a}^*,\vec{a}]$ for some $\vec{a}^* \in A$ it will be sufficient to prove that

$$\mathfrak{A} \models \psi^*[\vec{a}^*,\vec{a},\vec{c}] \quad \text{for some} \quad \vec{a}^*,\vec{c} \in A.$$

In the case where ψ^* is $0 < z_1$ the result is immediate. Now we consider case D) (b) above. We know that for some $\vec{d}$ we have $\mathfrak{B} \models \psi^*[\vec{b},\vec{a},\vec{d}]$.

Firstly we show that if $b \in B$, $a,a' \in A$ and:

(i) $b <_B a$ (or $a <_B b$) and there are finitely many elements in B between b and a

or (ii) $a <_B b <_B a'$ and there are finitely many elements in A between a and a'

then $b \in A$.

In the first case $\mathfrak{B} \models S[b,e_1]$ & ... & $S[e_n,a]$ for some $e_1,\ldots,e_n \in B$.

Since $T^* \vdash \forall x \exists! y\ S(x,y)$ & $\forall z(\neg z = 0 \to \exists! u\ S(u,z))$ and $S^{\mathfrak{A}} = S^{\mathfrak{B}} \cap (A \times A)$ it then follows that $b,e_1,\ldots,e_n,a \in A$. (ii) is proved in a similar way.

We must find $\vec{a}^*,\vec{c} \in A$ to correspond to $\vec{b},\vec{d} \in B$ such that the same order relations hold between corresponding elements and $\vec{a}$ and, further, when $\mathfrak{B} \models S[b_1,b_2]$ then $\mathfrak{A} \models S[a_1^*,a_2^*]$ (a_i^* corresponds to b_i).

If $b \in \vec{b}$ (or $\vec{d}$) and $b \in A$ then there is no problem. (Take b to correspond to b). Suppose $\vec{a} = a_1,\ldots,a_n$ where $a_1 < \ldots. < a_n$.

For those elements $b \in \vec{b},\vec{d}$ *not* in A one of the following holds:

(i) $0^A < b < a_1$ and there are infinitely many elements in A between 0^A and a_1;

(ii) $a_i < b < a_{i+1}$ and there are infinitely many elements in A between a_i and a_{i+1};

(iii) $a_n < b$ and there are infinitely many elements in B between a_n and b.

In each case it is clear that we can find $a^* \in A$ to correspond to $b \in B$ such that the order and successor relations are preserved. (In case (iii) it may happen that there are finitely many elements between a^* and a_n. We are only concerned that if $\mathfrak{B} \models S[a_i,b]$ then $\mathfrak{A} \models S[a_i,a^*]$ and *not* that if $\mathfrak{B} \models \neg S[a_i,b]$ then $\mathfrak{A} \models \neg S[a_i,a^*]$.)

This completes the proof that $\mathfrak{A} \models \psi[\vec{a}^*,\vec{a}]$ and hence by corollary 4.38, that T^* is complete.†

We now wish to show that T_{DIS} is complete.

Suppose $\sigma \in Sent(L)$ where L is the language of T_{DIS}. Since T^* is complete (and σ is a sentence in the language of T^*) either $T^* \vdash \sigma$ or $T^* \vdash \neg\sigma$. Without loss of generality suppose $\phi_1,\ldots,\phi_n = \sigma$ is a derivation of σ in T^*. The constant 0 and the predicate symbol S may occur in the derivation (although not in σ). However we replace an atomic subformula $A(t,0)$ in ϕ_i by $\exists y(A(t,y) \& \forall z(\neg z < y))$ and each atomic subformula $S(t_1,t_2)$ by $t_1 < t_2 \& \neg\exists z(t_1 < z \& z < t_2)$.

Let $\overline{\phi}_i$ denote this new formula of L. If ϕ_i is an axiom of T^* then $\overline{\phi}_i$ is a theorem of T_{DIS}. Hence $\overline{\sigma}(=\sigma)$ is a theorem of T_{DIS}. So, finally, we have shown that T_{DIS} is complete.

Exercise 4.12. (a) Show that the theory of discrete linear orderings with last but no first element is complete.

(b) Show that the theory of discrete linear orderings with first and last elements is not complete.

Exercise 4.13. T_1 is the theory of dense linear orderings with a first but no last element in the language with $=$, $<$. Show that T_1 is not model-complete but that if T_2 is the same theory in an extended language with $=$, $<$ and 0 then T_2 is model-complete.

A theory T has the *joint embedding property* (JEP) if given any two models of T, $\mathfrak{A}$, $\mathfrak{B}$ there is a model $\mathfrak{W}$ of T and embeddings $f: \mathfrak{A} \to \mathfrak{W}$, $g: \mathfrak{B} \to \mathfrak{W}$.

Exercise 4.14. Show that the following are equivalent:

(i) T has the JEP;

---

†A similar (but simpler) argument is used to show DLO is model complete. In this case the language does not have to be extended and ψ^* is a conjunction of formulae of the form $t < t'$.

(ii) for every $\phi, \psi \in \forall_1$ $Sent(L)$ if $T \vdash \phi \vee \psi$ then $T \vdash \phi$ or $T \vdash \psi$

(iii) for every $\phi, \psi \in \exists_1$ $Sent(L)$ if $T \cup \{\phi\}$, $T \cup \{\psi\}$ are consistent then $T \cup \{\phi \ \& \ \psi\}$ is consistent.

Exercise 4.15. Show that the theory of groups (in a language with $\cdot$, 1) has the JEP but that the theory of fields does not.

Exercise 4.16. Show that if T is model-complete and has the JEP then T is complete.

Exercise 4.17. Show that T is model complete iff given $\mathfrak{A}, \mathfrak{B} \models T$, $\mathfrak{A} \subseteq \mathfrak{B}$, $\phi(\vec{x}, y) \in Form(L)$, $\vec{a} \in A$, $b \in B$ if $\mathfrak{B} \models \phi[\vec{a}, b]$ then for some $a \in A$ $\mathfrak{B} \models \phi[\vec{a}, a]$.

Exercise 4.18. Show that if T is model-complete then T is axiomatizable by $\forall_2$ Sentences.

Exercise 4.19. Suppose $\mathfrak{A} \models T$. Show that the following are equivalent:

(i) $\mathfrak{A}$ is existentially closed over T;

(ii) if $\mathfrak{A} \models \phi[\vec{a}]$ for some $\phi(\vec{x}) \in \forall_1$ $Form(L)$, $\vec{a} \in A$ then $T \cup \Delta(\mathfrak{A}) \vdash \phi[\vec{a}]$;

(iii) if $\mathfrak{A} \models \phi[\vec{a}]$ for some $\phi(\vec{x}) \in \forall_1$ $Form(L)$, $\vec{a} \in A$ then there is $\theta(\vec{x}) \in \exists_1$ $Form(L)$ such that $\mathfrak{A} \models \theta[\vec{a}]$ and $T \models \forall \vec{x}(\theta(\vec{x}) \to \phi(\vec{x}))$.

Exercise 4.20. Suppose $\mathfrak{B}, \mathfrak{W} \models T$ such that $\mathfrak{B} \prec_1 \mathfrak{W}$ and $\mathfrak{W}$ is existentially closed over T. Show that $\mathfrak{B}$ is existentially closed over T.

A *chain* is a sequence of structures $\langle \mathfrak{A}_\alpha : \alpha < \beta \rangle$ of the same type such that for $\alpha_1 < \alpha_2 < \beta$ $\mathfrak{A}_{\alpha_1} \subseteq \mathfrak{A}_{\alpha_2}$. The *union of the chain* is the unique structure $\mathfrak{A}$ (of the same type) with domain $A = \bigcup_{\alpha < \beta} A_\alpha$ such that for $\alpha < \beta$ $\mathfrak{A}_\alpha \subseteq \mathfrak{A}$.

Exercise 4.21. Suppose $\langle \mathfrak{S}_\alpha : \alpha < \beta \rangle$ is a chain of structures such that for $\alpha_1 < \alpha_2 < \beta$ $\mathfrak{S}_{\alpha_1} \prec \mathfrak{S}_{\alpha_2}$. Show that the union of the chain is an elementary extension of each $\mathfrak{S}_\alpha$. Deduce that if T is model-complete then the union of any chain of models of T is a model of T.

Exercise 4.22. (Lindström's theorem 1964).
Suppose T is a theory in a countable language such that

(i) all models of T are infinite;
(ii) the union of any chain of models of T is a model of T.

Prove that T has an existentially closed model of each infinite cardinality κ.

If, further, T is κ-categorical for some infinite κ show that T is model-complete.

4.4. COMPLETENESS AND DECIDABILITY.

We close with some elementary observations on the consequences of completeness concerned with decidability. A comprehensive study requires knowledge of recursive function theory and so outline proofs only are given.

At the end of the first chapter examples were given of formulae ϕ for which the question: 'Is ϕ logically valid?' could be answered using a well defined method, a decision procedure. By the completeness theorem, for such a formula ϕ the question: 'is ϕ derivable (from the empty set)?' is also decidable. A mathematician is in general concerned not with logically valid formulae ϕ but with formulae ϕ valid in all models of some theory T, such as groups, fields, etc. Again appealing to the completeness theorem, the question is essentially: 'given ϕ in the language of T is it the case that $T \vdash \phi$?' The completeness of certain (countable) T has an important consequence bearing on this problem.

Definition 4.39. T is a consistent countable theory. T is *finitely (recursively) axiomatizable* if there is a finite (recursive) subset $T_0 \subseteq T$ such that for all $\phi \in T$, $T_0 \vdash \phi$.

Theorem 4.40. *If T is a complete recursively axiomatizable theory then T is recursive (decidable).*

Proof: We must show that there in a recursive procedure to determine of any sentence σ in the language of T whether or not $\sigma \in T$. Since T is complete we know that given σ either $\sigma \in T$ or $\neg\sigma \in T$. We let $\phi_1, \phi_2, \ldots$ be a recursive enumeration of the recursive set of axioms for T. At the nth stage we consider all those theorems derivable from $\phi_1, \ldots, \phi_n$ in at most n steps. This is a finite set. At some (finite) stage in this procedure either σ or $\neg\sigma$ will occur as a theorem. So, since T is consistent, we can determine if $\sigma \in T$. □

Exercise 4.23. Suppose T is a consistent theory with no infinite normal models. Show that T has only finitely many non-isomorphic models and that T is decidable.

Exercise 4.24. Suppose T is a recursively axiomatizable theory with only finitely many distinct theories T_i $(1 \le i \le m)$ such that $T \subseteq T_i$. Show that all the T_i and T are decidable.

FURTHER READING

We finish with a few suggestions for further reading on model theory for the interested reader.

Chang and Keisler (1973) is a very comprehensive text on model theory that must now count as essential reading for a serious student of the student subject.

Robinson, whose work on model theory during the fifties and sixties is fundamental in the subject, wrote several texts. The most recent edition (1965) includes the material of our Chapter 4 in greater detail together with other important topics.

Bell and Slomson (1969) approaches the subject via Boolean algebras and the ultraproduct construction. (It includes the Rasiowa-Sikorski proof of the completeness theorem using Boolean algebras.)

Sacks' (1972) elegant and concise book includes a proof of Morley's result (4.16) and other related theorems.

The selection we have given is in no way comprehensive. Each of the above texts contains a lengthy bibliography with still more suggestions of books and papers on model theory that should provide sufficient stimulus for the keenest student.

REFERENCES

Aczel, P. (1973). *Infinitary logic and the Barwise compactness theorem*, in: The Proceedings of the Bertrand Russell Memorial Logic Conference, Denmark 1971, J. L. Bell, J. Cole, G. Priest and A. B. Slomson, Eds. 234-277 (B. Russell Logic Conference, Leeds).

Bell, J.L. and Slomson, A.B. (1969). *Models and ultraproducts*, (North-Holland, Amsterdam).

Beth, E.W. (1955). *Semantic entailment and formal derivability*, Mededelingen der Koninklijke Nederlandse Akademie van wetenschappen, afd. letterkunde, n.s., 18 309-342.

Cantor, G. (1895). *Beiträge zur Begründung der transfiniten Mengenlehre*, Math. Ann. 46, 481-512 (English translation in: Contributions to the Founding of the Theory of Transfinite Numbers (Dover, New York 1915)).

Chang, C.C. and Keisler, H.J. (1973). *Model theory*, (North-Holland, Amsterdam).

Church, A. (1936). *A note on the Entscheidungsproblem*, J. Symb. Logic 1 40-41, correction ibid. 101-102.

Enderton, H.B. (1972). *A mathematical introduction to logic*, Academic Press, New York).

Gandy, R.O. (1973). *'Structure' in mathematics*, in Structuralism: an Introduction, D. Robey, Ed. (Clarendon Press, Oxford).

Gödel, K. (1930). *Die Vollständigkeit der Axiome des logischen Funktionenkalküls*, Monatsh. Math. Phys. 37, 349-360 (English translation in: From Frege to Gödel, J. van Heijenoort, Ed. 582-591 (Harvard 1967)).

(1931). *Über formal unentscheidbare Sätze der Principia Mathematica und verwandter Systeme I*, Monatsch. Math.Phys.38, 173-198 (English translation in: From Frege to Gödel. J. van Heijenoort, Ed.596-616 (Harvard 1967)).

Halmos, P.R. (1960). *Naive set theory*, Van Nostrand,Princeton). (N.B. Latest edition published by Springer-Verlag,Berlin).

Henkin, L.A. (1949). *The completeness of the first order functional calculus*, J. Symb. Logic 14, 159-166.

Henkin, L.A. (1950). *Completeness in the theory of types*, J. Symb. Logic 15, 81-91.

Herbrand, J. (1930). *Recherches sur la théorie de la démonstration*, Trav. Soc. Sci. Lett. Varsovie, Classe III 33 (English translation of Ch.5 in: From Frege to Gödel, J. van Heijenoort, Ed., 525-581 (Harvard 1967).)

Kleene, S.C. (1952). *Introduction to metamathematics*, (Van Nostrand, Princeton).

Kreisel, G. and Krivine, J.L. (1967). *Elements of mathematical logic: model theory*, (North-Holland, Amsterdam).

Lemmon, E.J. (1965). *Beginning logic*, (Nelson).

Lindström, P. (1964). *On model completeness*, Theoria (Lund) 30, 183-196.

Löwenheim, L. (1915). *Über Moglichkeiten in Relativkalkül*, Math.Ann. 76, 447-470 (English translation in: From Frege to Gödel, J. van Heijenoort, Ed., 228-251 (Harvard 1967)).

Malcev, A. (1936). *Untersuchungen aus dem Gebiete der mathematischen Logik*, Rec. Math. N.S.1, 323-336.

Mendelson, E. (1964). *Introduction to mathematical logic*, (Van Nostrand, Princeton).

Morley, M. (1965). *Categoricity in power*, Trans. Am. Math. Soc. 114, 514-538.

Rasiowa, H. and Sikorski, R. (1950). *A proof of the completeness theorem of Gödel*, Fund. Math. 37, 193-200.

Robinson, A. (1956). *Complete theories*, (North-Holland, Amsterdam).

(1965). *Introduction to model theory and to the metamathematics of algebra*, (North-Holland, Amsterdam).

Rogers, H. (1967). *Theory of recursive functions and effective computability*, (McGraw-Hill, New York).

Sacks, G. (1972). *Saturated model theory*, (Benjamin).

Schoenfield, J.R. (1967). *Mathematical Logic*, (Addison-Wesley, Reading, Mass.).

Simmons, H. (1972). *Existentially closed structures*, J.Symb. Logic 37, 293-310.

Skolem, T. (1920). *Logisch-kombinatorische Untersuchungen über die Erfüllbarkeit oder Beweisbarkeit mathematischer Sätze nebst einem Theoreme über dichte Mengen*, Skrifter utgitt av Videnskapsselskapet i Kristiania, I. Mat. Naturv. Kl. 4. (English translation in: From Frege to Gödel, J. van Heijenoort, Ed., 252-263 (Harvard 1967)).

(1922). *Einige Bernerkungen zur axiomatischen Begründung der Mengenlehre*, Mathematikakongressen i Helsingfors den 4-7 Juli 1922, Den femte skandanaviska matematikakongressen, Redogörelse (Akademiska Bokhandeln, Helsinki 1923) (English translation in: From Frege to Gödel, J. van Heijenoort, Ed., 290-301 (Harvard 1967)).

Steinitz, E. (1910). *Algebraische Theorie der Körper*, J. Reine Angew. Math. 137, 167-309.

Tarski, A. (1935). *Der Wahrheitzbegriff in den formalisierten Sprachen*, Studia Philos. (Warsaw) 1, 261-405 (English translation in: Logic, Semantics and Meta-mathematics 152-278 (Clarendon Press, Oxford 1956)).

Tarski, A. and Vaught, R. (1957). *Arithmetical extensions of relational systems*, Compositio Math. 13, 81-102.

Vaught, R. (1954). *Applications of the Löwenheim-Skolem-Tarski theorem to problems of completeness and decidability*, Koninkl. Ned. Akad. Wetensch. Proc. Ser. A 57 (Indag. Math 16 467-472).

Wang, H. (1952). *Logic of many sorted theories*. J. Symb. Logic 17, 105-116.

INDEX